Bioethics, Public Moral Argument, and Social Responsibility

Routledge Annals of Bioethics

Series Editors: MARK J. CHERRY, *St. Edward's University, USA*
ANA SMITH ILTIS, *Wake Forest University, USA*

Bioethics, Public Moral Argument, and Social Responsibility

Edited by Nancy M. P. King and Michael J. Hyde

Routledge
Taylor & Francis Group
NEW YORK   LONDON

First published 2012
by Routledge
711 Third Avenue, New York, NY 10017

Simultaneously published in the UK
by Routledge
2 Park Square, Milton Park, Abingdon, Oxon OX14 4RN

*Routledge is an imprint of the Taylor & Francis Group,
an informa business*

© 2012 Taylor & Francis
The right of the editor to be identified as the author of the editorial material, and of the authors for their individual chapters, has been asserted in
accordance with sections 77 and 78 of the Copyright, Designs and Patents
Act 1988

Typeset in Sabon by IBT Global.

Library of Congress Cataloging-in-Publication Data
    Bioethics, public moral argument, and social responsibility / edited by
Nancy M.P. King and Michael J Hyde.
       p. cm. — (Routledge annals of bioethics)
    Includes bibliographical references and index.
    ISBN 978-0-415-89855-3 (hardback)
    1.   Medical ethics—Philosophy. 2.   Bioethics—Philosophy.
    3.   Biotechnology—Philosophy. I.   King, Nancy M. P. II.   Hyde,
Michael J., 1950–
    R725.5.B523 2011
    174.201—dc23
    2011016378

ISBN13: 978-0-415-89855-3 (hbk)
ISBN13: 978-0-203-63051-8 (ebk)

Contents

PART III:
The Media, the Public, and the Person

Acknowledgments

This volume is made possible by several funding sources at Wake Forest University: the Ethics, Leadership, and Civic Responsibility Fund, the Provost's Fund, the Archie Fund, the Center for Bioethics, Health, and Society, and the Department of Communication. Much valued support also came from our colleagues, staffs, and students. Special thanks are owed to Jill Tiefenthaler, Beth Hutchens, Susan Harris, and Michael Tennison. We are particularly grateful to Wake Forest University, the Master of Arts in Bioethics Program, and Wake Forest Baptist Medical Center for supporting the growth of the scholarly relationship between bioethics and communication. Finally, our utmost gratitude is reserved for the superb contributions of our distinguished colleagues who have authored chapters in this volume.

Editors' Introduction

Nancy M. P. King and Michael J. Hyde

THE CALL TO CONVERSATION

In November 2001, the newly formed President's Council on Bioethics (PCB) was charged with deliberating about the benefits and burdens of biotechnology and then publishing these deliberations as a way to "spark and inform public debate." Then-President Bush appointed the noted physician and conservative bioethics scholar Leon Kass to chair the PCB, and charged it, among other things, "to undertake fundamental inquiry into the human and moral significance of developments in biomedical and behavioral science and technology; to explore specific ethical and policy questions related to these developments; to provide a forum for a national discussion of bioethical issues; [and] to facilitate a greater understanding of bioethical issues." (Executive Order 13237). The PCB's charge continued: "The Council shall strive to develop a deep and comprehensive understanding of the issues that it considers. In pursuit of this goal, the Council shall be guided by the need to articulate fully the complex and often competing moral positions on any given issue, rather than by an overriding concern to find consensus. The Council may therefore choose to proceed by offering a variety of views on a particular issue, rather than attempt to reach a single consensus position."

This mandate—to consider deeply and discuss fully—is unique in the long history of presidential bioethics bodies. When President Obama's new Presidential Commission for the Study of Bioethical Issues began its work in July 2010, a return to the more pragmatic mandate of prior commissions had already been announced: "The Commission shall pursue its work with the goal of identifying and promoting policies and practices that ensure scientific research, healthcare delivery, and technological innovation are conducted in an ethically responsible manner." (Executive Order 13521) Earlier presidential bioethics bodies often succeeded in reaching consensus about policy directions, but on more than one occasion members reported agreeing on conclusions while disagreeing profoundly on the reasoning used to reach them. This type of policy consensus is certainly sufficient for some purposes, but it is not necessarily supportive of critical reflection on questions and issues in need of public input. For this reason, despite its appearance to some as a "debating society," the now-disbanded PCB modeled a practice of public discourse that others may come to miss (Hyde 2008; 2010, pp. 211–41).

Through this volume we seek to challenge scholars in bioethics and communication to promote meaningful public discussion of how science and medicine affect and should affect our lives. We've gathered here a range of voices to explore the role of democratically-oriented argument in promoting public understanding and discussion of the benefits and burdens of biotechnological progress, with the goal of developing and applying a collective wisdom to the trajectory of modern biomedical science. The communication and rhetorical practice of such public moral argument requires experts from the sciences and the humanities to step beyond their respective disciplinary boundaries and assume the ethical responsibility of translating their expertise into forms that help promote public conversation about important matters of concern. The essays collected and organized here are our first foray into the development of a collective voice, made possible by the workings of public moral argument. The last essay, authored by two nationally recognized college debate scholars and teachers who reviewed the essays, offers a critical analysis of what their colleagues did and did not say about the nature of "the public" and its role in moral argument. Public moral argument is certainly called for in today's ongoing biotechnology debate. Our volume is a response to this call.

PUBLIC MORAL ARGUMENT[1]

Scholars of public moral argument make their living by studying the symbolic capacities of human beings, especially as these capacities show themselves in situations that call for the production of discourse as a means for coming to terms with the matters at hand. This call emerges from some perceived exigency, or what Bitzer (1968) terms "an imperfection marked by urgency." An exigency is rhetorical when it "invites the assistance of discourse" as a way of implementing change that can result in some "positive modification" of the imperfection (pp. 6–7). When language is used to respond to a rhetorical exigency, its technological nature becomes obvious; in these situations discourse is being employed as a tool, an instrument, a means to an end. In this respect, human language can even be described as a biotechnology.

Research in the ethics of health communication examines how language is used as a technology—in particular, how it informs the interpersonal dynamic between physicians and those whom they are obliged to serve. The basic goal of this research is to discover how the communicative and rhetorical competence of the involved parties (e.g., their ability to construct informative narratives) can be perfected in order to produce measurable, effective, and good health-care outcomes. Such research, however, is not restricted to the interpersonal settings of the physician/patient/family encounter. The well-being of the body politic of democracy requires that a process like informed consent transcend the institutional boundaries of the medical establishment in order to educate the citizenry about biotechnological progress. This educational process encourages the production of whatever public moral argument may be necessary for understanding and dealing with both the benefits and burdens associated with this progress and its perfectionist impulse.

Scientific medicine was born with the help of public moral argument. Trained by the Sophists of their day, Hippocratic physicians involved themselves in this communication and rhetorical process when defining and defending their *techne* during public debates and while treating patients. For these first men of scientific medicine, the biotechnology of language served the important purpose of calling into being a "medical public" that, owing to its new scientific education, could stand with the Hippocratic physicians in their initial fight against traveling sophistic lecturers and those quack doctors whose practice still admitted the use of magical charms (Edelstein 1987, pp. 87–110; Laín Entralgo 1970, pp. 139–170; Frede 1987, pp. 232–239; Jonsen 1990, pp. 8–9; Hyde 2001, pp. 124–129).

Plato commended this rhetoric of science in his *Laws* (IV, 720c-e). Hippocratic physicians employed it, however, so as to be done with it. As noted in the Hippocratic text *Decorum*, the wisdom that these healers possess and that they must constantly seek as their first priority makes them "the equal of a god. Between wisdom and medicine there is no gulf fixed" (Jones 1923a, V). The point is put another way in the Hippocratic *Law*: "There are in fact two things, science and opinion; the former begets knowledge, the latter ignorance" (Jones 1923b, IV).

The birth of scientific medicine sharpened this long-standing dispute between the arts and humanities and the sciences over the degree of respect that each owes the other. The biotechnology of utmost importance to medicine today is arguably not the word, but rather those other tools that enhance the scientific capacity of medicine to prevent, treat, or cure a host of life-threatening illnesses: tools like immunization against childhood virus diseases, antibiotics for bacterial infections, surgical procedures for organ transplantation, life-sustaining ventilators, respirators, and dialysis machines, cancer chemotherapy, genetic engineering, and embryonic stem cell research. The view of medical science as exclusively to save, enhance, and extend life presents a false but persistent dichotomy: that patients must choose either "the doctor who will cure you or the one who will hold your hand and talk to you." There is need for both—and even as biotechnology advances, it is becoming harder to believe that either can happen without the other. Human beings desire both cure and care (Brody 2009; Hyde 2006, pp. 1–10).

The goals of biomedical technology thus incorporate a key public concern about the meaning of being human. When the case of Terri Schiavo first made news, many bioethics scholars had the initial reaction: "But we've already solved that problem!" Readers will recall that Terri Schiavo was a young woman who unexpectedly collapsed, and after a period of time was diagnosed as being in a permanent vegetative state. She had no formal advance directive. After an aggressive search for means to restore her awareness, cognition, and dignity, her husband concluded that she could not recover, but her parents concluded that she had been misdiagnosed and was treatable. When the legal battles began over who spoke for Terri, she joined the short list of young women whose medical fates have shaped American views about the life worth living: first, in the 1970s, Karen Ann Quinlan; then, in 1990, Nancy Cruzan; and now, in the 21st century, Terri Schiavo.

Why did bioethics scholars think that the Terri Schiavo problem had been solved by Karen Quinlan and Nancy Cruzan? By enabling the long-term survival of patients in various states of permanent unconsciousness, technology had created both a new diagnosis and a new dilemma. The stories of Karen Quinlan and Nancy Cruzan spurred profound legal and policy changes to address the new diagnosis and the role of families in making health care decisions for adults newly unable to decide for themselves. Yet as Art Frank (1995, 2004) has observed, when patients and families find themselves facing this dilemma, it is always new for each of them. Therefore, the value placed on human life and human dignity in that diminished state must be adjudicated anew, in every new instance, through respectful moral discourse. This discourse often involves a great many stakeholders seeking a voice: not only the patient and the patient's legally authorized decision-maker(s), family, and friends, but also the health care team, the institution, the state, advocacy groups, scientists, and scholars with different perspectives, health insurers paying the bills, and more.

Public moral argument is thus called for to elucidate society's role, both at the end of life and about the ends of life. That role is messy, disputed, limited—and essential. Much discussion of biotechnological advances in the U.S. rests on the rights of individuals to make autonomous choices and on societal decisions not to interfere with willing buyers and sellers. We might simply acknowledge that this (admittedly incomplete) laissez-faire position is the role that society has chosen, that it has particular consequences, and that we could, but need not, choose otherwise. But the current, intensifying democratic debate about related matters, such as health insurance reform, has begun to broaden our public vision, to include awareness of cost, a sense of collective responsibility to help others, and the need to work together to set limits we can live with. Science and society have built an exceedingly and increasingly complex community around biomedical technology. Scholars, scientists, policymakers, and the public all therefore need to be able to talk together in this community. This important effort brings together the enterprises of bioethics and communication ethics, with their shared interests in the health of human beings and in the social, political, and technical ways of using language to affect health and health care; hence the essays in this volume.

THE ESSAYS

These essays cluster around three complementary themes, which evolve as the reader progresses through the volume. Our organization of chapters into three parts—Public Moral Argument and Social Responsibility, Moral Relationships and Responsibilities, and the Media, the Public, and the Person—represents our best attempt to trace these themes. However, thoughtful readers will readily recognize ways in which the themes are braided together throughout the chapters, reflecting their salience both for scholars and for public discourse.

One theme, explored perhaps most explicitly by Moreno, Zarefsky, and Coughlin et al. in Part I, examines moral language and moral relationships: that is, the means by which moral engagement is fostered in American society. There are necessary tensions between moral authority and argument and social and political decision-making. Both advocacy and consensus-seeking test the taut balances that democracy requires. Much more attention should be paid to the complex role of uncertainty and fallibility in the face of the need for decision and action. Existing models of public moral argument need to illustrate and teach responsible advocacy and decision-making under uncertainty, in order to model productive relationships between scholars and society. New discourse models may be needed as well, to ensure that democratic decision-making can flourish in a marketplace shaped as much by technology as by ideas.

A second theme addresses the nature of selfhood and moral agency. This is the theme captured in Part II by Churchill, Dresser, Parrott, and Juengst. The language used in discussions of "human nature" and its relationship to critical concepts in biomedical technology, like enhancement and genetics, stems from social, cultural, and religious understandings that merit careful examination, for several reasons. They may be based on outmoded or discriminatory views that should be uncovered and cautioned against. Alternatively, they may reflect rich, nuanced, flexible, and pragmatic perspectives that can expand our collective vision and therefore should be emulated and promulgated. Listening to how people actually talk, and learning how people actually behave, in light of the new knowledge about ourselves that biotechnological progress can provide are essential components of responsible genetic science. In other words, paying attention to what we say, how we think, feel, and act, helps us understand who we are. If health communicators are to play a meaningful role in helping the public make use of information about a set of critical issues—the ways that humans respond to technology, understanding the genetic contribution to health and illness, or the effects of treatment or enhancement on the sense of self—then discourse must be mutual and multidirectional.

A third theme focuses on moral responsibility in public discourse. As elucidated by Condit, Giles and Krcmar, Lundberg and Smith, and most provocatively by Elliott, the scholar's responsibility lies not only in calling others to account (whether through critiques of manipulative media like advertising or by highlighting the need to link policy meaningfully to accounts of the personal and the public) but also in making the hard choices and taking the risks that can accompany this essential public role. Thus it is essential to consider the responsibility of the individual scholar to address ethical issues that arise close to home, even when they can disturb scholarly distance and complacency; the responsibility of scientists, bioethics scholars and practitioners, and journalists to "get it right" as teachers of the general public; and the responsibility of respectful engagement, even when forging genuinely responsive relationships requires making time and taking risks. Challenging the scholar's traditional role of careful, dispassionate researcher and teacher goes to the heart of bioethics, asking that those who preach ethical behavior must also practice it, in every aspect of their professional lives. What this means for the social role of bioethics remains to be discovered—or, rather, created—by the writers and readers of this volume and others like it.

HEALTH CARE AND MORAL DISCOURSE TODAY

The need to consider carefully the meaning of responsible public moral argument—and the responsibility to achieve it—could hardly be more pressing than it is today. Moral argument and moral relationships are increasingly articulated not only in words, but in the images, technologies, and settings by which words are framed and delivered. How each of us uses and responds to data and devices, and to the people we encounter and affect by and through them, are key concerns in public health, health care, and health research—and in our social engagement with all three. This is the stuff of bioethics: not merely a set of issues, topics, and cases, but the broadest and deepest consideration of the human implications of the life sciences, beginning—and ending—in our collective and continuing conversation.

NOTE

1. In this section we draw on Hyde & King 2010.

REFERENCES

Bitzer, L. F. (1968). The rhetorical situation. *Philosophy and Rhetoric* 1: 1–14.
Brody, H. (2009). *The future of bioethics.* New York: Oxford University Press.
Edelstein, L. (1987). *Ancient medicine.* Trans. C. L. Temkin. Ed. O. Temkin and C. L. Temkin. Baltimore: John Hopkins University Press.
Executive Order 13237, November 28, 2001.

Executive Order 13521, November 24, 2009.

Frank, A. (1995) *The wounded storyteller: Body, illness, and ethics*. Chicago: University of Chicago Press.

Frank, A. (2004) *The renewal of generosity: Illness, medicine, and how to live*. Chicago: University of Chicago Press.

Frede, M. (1987). *Essays in ancient philosophy*. Minneapolis: University of Minnesota Press.

Hyde, M. J. (2001). *The call of conscience: Heidegger and Levinas, rhetoric and the euthanasia debate*. Columbia: University of South Carolina Press.

Hyde, M. J. (2006). *The Life-Giving Gift of Acknowledgment*. West Lafayette, IN: Purdue University Press.

Hyde, M. J. (2008). *Perfection, postmodern culture, and the biotechnology debate*. New York: Pearson/Allyn & Bacon.

Hyde, M. J. (2010). *Perfection: Coming to terms with being human*. Waco, TX: Baylor University Press.

Hyde, M. J. & King, N. M. P. (2010). Communication ethics and bioethics: an interface. *Review Of Communication* 10: 156–171.

Jones, W. H. S. (1923a). *Hippocrates (Decorum)*, vol. 2. London: William Heinemann; New York: G. P. Putnam's Sons.

Jones, W. H. S. (1923b). *Hippocrates (Law)*, vol. 2. London: William Heinemann; New York: G. P. Putnam's Sons.

Jonsen, A. R. (1990). *The new medicine and the old ethics*. Cambridge, MA: Harvard University Press.

Laín Entralgo, P. (1970). *The therapy of the world in classical antiquity*. Trans. L. J. Rather and J. M. Sharp. New Haven: Yale University Press.

Plato (1961). *Laws*. Trans. A. E. Taylor. In E. Hamilton & H. Cairns (Eds.), *Plato: The collected dialogues*. Princeton: Princeton University Press.

Part I
Public Moral Argument and Social Responsibility

1 Arguing About Values
The Problem of Public Moral Argument

David Zarefsky

I claim expertise neither in medicine nor in ethics, but I do study public argument. And if it is true that our scientific and medical knowledge have outpaced our ethical understanding, then even more has our ethical understanding outpaced our ability to argue effectively about moral or ethical issues. This condition is especially serious because public argument is the means by which a democratic society comes to judgment and decision about matters of controversy.

THE TENSION BETWEEN DEMOCRACY AND MORALITY

An explanation of our predicament must begin with an understanding of the tension between democracy and moral argument. For a working definition of democracy, I'll use the one Abraham Lincoln put forth when he summoned Congress into special session following the attack on Fort Sumter: "a government of the people, by the same people" (Lincoln 1861/1953b)—a phrase that prefigures the Gettysburg Address. The key idea is that, in a democracy, sovereignty resides with the people. They delegate power to their leaders, whom they expect to act on their behalf and whom they hold accountable. For the secular, sovereignty resides in the people by virtue of natural rights; for the religious, as a gift of God.

If sovereignty resides in the people, three corollaries follow (Zarefsky 2008). One is political equality. It is not that people in fact are equal in power and influence, but that decision-making authority is allocated on a per capita basis, not on the basis of wealth, race, gender, religion, heredity, or intelligence. A second corollary is majority rule. People will not all think alike, yet decisions must be made in the face of uncertainty. If each has equal access to decision-making authority, then it follows that decisions must be made by the weight of greater numbers. And the third corollary is minority rights. Even though they do not prevail, members of the minority retain their legitimacy and sovereignty, and they could become the majority another day. Democracy is like an ongoing conversation; there are no final victories.

A democratic society is grounded in the assumption of human fallibility (Thorson 1962). We commit ourselves to certain beliefs; we think we are right; but we cannot know *for sure*. This human imperfection may be the result of unfinished evolution or of original sin, but the fact is that we could be wrong. For our ideas to be widely accepted, therefore, we must rely not on their inherent truth but on the free assent of others. And when their judgment is that our ideas are wrong, society will abandon those ideas and adopt others. There are no final, absolute victories. The virtue of democracy is that it permits and encourages the correction of error.

But there is a tradition of discourse that challenges these assumptions; it is the moral voice. It traces back to the prophets of the Hebrew Bible. They did not seek the assent of their audiences, or if they did, they went about it in a very strange way. Excoriation was their mode of operations; they called listeners to account for their misdeeds and challenged them to repent lest Divine punishment be even more severe. They had no doubt that they were right. They knew *for sure* because they were not expressing their own ideas. They were merely messengers transmitting the word of God—"thus saith the Lord." The prophetic voice was not stilled when the Biblical canon was closed. Even today, some participants in moral controversies will claim absolute certainty resulting from their access to God's Word.

Democracy presumes fallibility; prophecy presumes certainty. Yet it is an even more complicated tension than that. Democracy is not a purely procedural system, and it is the prophetic voice that enables a democracy to evolve. The abolitionist movement of the 19th century and the civil rights movement of the 20th century were inspired by moral appeals. The fundamental evil of slavery was that it denied the slave the dignity inherent in personhood, and it thereby degraded the dignity of the master as well. This was not a contingent proposition; the abolitionists knew it *for sure*, and decades of controversy and the circumstances of war convinced the vast majority of Americans that they were right. The civil rights movement followed the premise that we are all God's children—a premise about which its advocates were *certain*—and argued to the conclusion that racial discrimination, with its assumption of superiority and inferiority, had no place in American life. Racism has not disappeared, but over the past 60 years we have come to accept that officially-sanctioned discrimination is wrong.

One finds the prophetic moral voice in many other controversies—in calls to extend rights and liberties, and in calls to restrict them. It has figured prominently in the movements for women's rights and, more recently, gay rights; it also has figured prominently in the movements for prohibition and for sexual abstinence before marriage. The paradox is that the prophetic voice is at odds with democracy and yet may be essential in enabling a democracy to advance

It should not be surprising, then, that moral conflicts are particularly difficult. Nor is this anything new. One hundred and fifty years ago, perhaps the greatest champion of democracy (with both an upper-case and a lower-case "d") was Stephen A. Douglas. Slavery is a complex moral issue, he said, and it is not given to us to know which side is right. So rather than legislate for or against slavery in the territories, let the decision be made by those who go there to live. When a group of Chicago clergymen chastised him for moral obtuseness, he rebuked them, insisting that they had no special authority to speak on the matter (Douglas 1854/1961). On the other hand, John Brown knew *for sure* that slavery was an evil. It was beyond doubt or argument; his conscience demanded of him existential acts, that he do what he could to purge the nation of its sins. Equally convinced, however, was William Lowndes Yancey, a Southern fire-eater who knew *for sure,* because it was in the Bible, that slavery was a positive good and therefore that Congress must act affirmatively to protect the property rights of the slaveholders by enacting a slave code for the territories. With hindsight we can say that the genius of Abraham Lincoln was that he fused the prophetic and the pragmatic. He began with the premise that slavery was wrong, just as John Brown and William Lloyd Garrison did, but unlike them he reached the prudent conclusion that it should be *contained*—a position that enabled "strange, discordant, and even hostile elements" (Lincoln 1858/1953a) to coalesce under the banner of the Republican party.

The moral issues of our own time—issues such as abortion, cloning, stem cell research, gay marriage, and end-of-life decisions—are no less complex than the slavery issue was for our forebears. The experience of the slavery issue also suggests how we need to proceed with our own disputes. We need to argue them out, seeking the assent of our fellow human beings.

Arguing about values is difficult. Even *acknowledging* value conflicts is hard. We may avoid them because we think they don't affect us, or because we don't want to offend others, or because we don't want to pass judgment and would rather "live and let live." We don't want to *argue* about them because that seems like bickering and fighting, or because we don't want to risk exposing our beliefs to scrutiny, or because we imagine that there is no way to resolve a dispute: you have your values and I have mine, and that is that.

But if we do not engage our values in argument, we cannot make decisions democratically. We must either rely on some kind of force—the coercion of military power, the weight of authority, or the threat of reprisal—or we must settle for pure relativism, according to which no one value is preferable to any other (Booth 1974). I may value freedom and you may value tyranny, and there is no way to choose between us. The history of the last century is littered with object lessons suggesting that we must not settle for these alternatives.

So let us attempt the task of arguing about values, engaging our moral judgments within the assumptions of a democratic society. We must recognize first that virtually all arguing about values is case-based. We will not get very far if we try to resolve our disputes in the abstract. In an essay for the President's Council on Bioethics, Adam Schulman notes that human dignity might be grounded in our higher mental capacities, or in the equality of all persons, or in individual autonomy and choice (Schulman 2008). In the abstract, most of us believe in *all* of those values, and yet in a particular situation they can lead to different, even incompatible, outcomes. So we have to make value judgments by arguing for the applicability of one or another value to the specific case. This involves the ancient faculty of prudence, or what the Greeks called *phronesis,* practical wisdom. It is not conclusive, nor final, nor generalizable. The same methods and materials are available to advocates on both sides of the dispute.

HOW WE ARGUE ABOUT VALUES

Levels of Argument

Arguments about values occur on two different levels. Sometimes the point of the argument is to determine that something is a value in its own right. In these cases the value is the claim to be established. The claim is defended by reasons that an audience would take to be justification for it, as well as by warrants derived from other values that the audience accepts. For example, one might defend the claim that reducing our carbon footprint is a moral obligation. Reasons might include evidence that we are depleting the world's natural resources and a warrant derived from other values might be that we have a stewardship responsibility to care for the earth. If the audience accepted the warrant and was convinced by the evidence, the combination of warrant and evidence would establish the obligation to reduce our carbon footprint. The advocate for this claim will want to use warrants that the audience is known to accept. If the warrant is not accepted, then it too will need to be established as a claim, and that would require an ancillary argument. Stewardship responsibilities, for example, could be warranted both by an appeal to justice and by their acknowledgment in the Bible. In theory, the search for warrants acceptable to the audience could produce an infinite regress, but it is highly unlikely that there will be *no* commonly accepted values. Abandoning the effort to find common values that can warrant other values should be the arguer's last resort.

A variation on this approach to arguing about values is the argument *a fortiori*. This is an argument about more and less. It suggests that the greater implies the lesser (or vice versa, as the case may be). If we have the responsibility to take care of the earth for the sake of future generalizations, then even more do we have the (subsidiary) responsibility to reduce harmful pollution from carbon emissions, which is one of the threats to the

future of the earth. Acceptance of the greater value should imply acceptance of the lesser value which it subsumes.

More common, however, is the second kind of value argumentation:
defending a choice between or among competing values. For example, in
the abstract we may value both telling the truth and showing empathy
and concern for others. But we are confronted with a practical situation
in which we must choose between these values. In a conversation with a
friend, should we tell the person what we honestly think about his or her
spouse, thereby being faithful to the value of truth-telling, or should we
tell a "white lie" in order to show empathy and concern for the friend and
the relationship? The answer will vary with the specific circumstances,
but in any given case we must be able to argue that one value should be
preferred over the other. Like this example, conflicts typically involve two
values that are good in themselves but may be incompatible in a specific
case, such as the conflict between liberty and equality. In principle we
support both of these values, yet each recedes as we maximize the other.
Sometimes there is a compromise tradeoff, but sometimes we want to
argue directly for the prominence of one over the other. When that is our
goal, how do we pursue it?

Strategies of Argument

First, we can argue that one value subsumes the other. By choosing one we
actually could enhance both. For instance, the controversy over whether to
undertake heroic measures to resuscitate patients believed to be terminally
ill can be understood as a conflict between the values of life and the quality
of life. Advocates on one side may say that valuing life is to be preferred
because life is a necessary condition for the quality of life; there is no point
in considering the quality of life after the patient has died. Conversely, however, one might maintain that the quality of life is precisely what makes life
meaningful and distinguishes it from mere existence.

Second, we might try to establish that pursuing one value yields a comparative benefit over pursuing the other. In considering priorities for public
spending, one advocate might contend that spending on education will be
an investment in the future; another might reply that spending on prisons
will assure our security in the present. Funds are limited and it is not possible to direct significant resources to both. Then the dispute will turn on
the question of whether greater benefit is achieved by focusing on the needs
of the future or of the present.

Third, we might argue for one value over another on the basis that it
has a greater likelihood of attainment. If we can achieve one value while
the other remains speculative, then it would seem reasonable to pursue the
one that could be obtained rather than risking the loss of both. An example
might be the vexing philosophical problem of whether justice should be
preferred over happiness, or vice versa. One might prefer the value of justice

on the grounds that it can be achieved in this world whereas true happiness can be achieved only in the next. Alternatively, one might maintain that one should pursue happiness because it is a state of mind, subject to our own control, whereas achieving justice depends upon the actions of others as well.

Fourth, we could argue that one value is preferred over another because it is a better means to a shared goal. In this case there is agreement on the terminal value to be sought but disagreement over the instrumental values that promote it. Virtually all parents, for example, want their children to grow into mature adults, but there is considerable disagreement about the values that will lead to that goal. One advocate might defend the value of autonomy, saying that giving children latitude to make many of their own decisions will provide experience in responsible decision-making that is a hallmark of maturity. Another might maintain that close supervision and direction is a better path to the goal, because the child who practices desirable behavior under parental supervision will develop a habit of it and hence will be more likely to behave appropriately on his or her own. The advocates would exchange reasons for believing that the instrumental values they support will be more likely to achieve the commonly-held terminal value.

Fifth, we might propose that one value is better supported by authoritative sources than is the other. This approach presumes that both advocates accept the authority of the source. For example, we might imagine two religious people arguing about the extent of human responsibility for the environment. The advocate who believes that this is a low priority might maintain that the world exists for human use, citing the Biblical admonition that humankind fill up the earth and subdue it. The other, who thinks that we must preserve the earth for future generations and that attending to this responsibility is a high priority, might cite the Biblical admonition to take care of the earth, claiming that we are stewards but that the earth does not belong to us. This can be a productive argument because both advocates accept the authority of the Bible. The question then is which of the competing Biblical texts more clearly applies to the case at hand. On the other hand, if one advocate regarded the Bible as a guide to conduct and the other regarded it only as an interesting narrative, then a prior argument would be needed about whether the Bible should be regarded as an authoritative source, and if not, what other source should be considered authoritative. Or if the arguers rely on different sources, each of which could lay claim to authority (based on experience, training, or previous judgment, for example), then it will be necessary to determine in the given case which source can lay the greater claim.

Sixth, we might appeal to what in rhetoric is referred to as the locus of the irreparable (Perelman & Olbrechts-Tyteca 1958/1969). This is an argument suggesting that if one choice is made (or not made, as the case may be), the consequences will be irreversible; we will be past the point of no return. The underlying assumption ordinarily is that preserving options is better than losing them. For example, an advocate might prioritize the value of energy conservation over energy use by noting that if we exhaust the earth's fossil fuels, we cannot replace them. Since we cannot know how quickly alternative fuels can be made available, it makes sense to slow the rate at which we deplete fossil fuels. On the other hand, an opposing advocate might argue that the current use of fossil fuels is essential to sustain economic growth, and without continued economic growth we not only will be unable to meet current social needs but also will lack the capital investment necessary to develop alternative fuel technologies. For one advocate, then, the locus of the irreparable is grounds for conserving fossil fuels while for another advocate it is grounds for continuing to use them.

These six broad patterns hardly exhaust the ways in which we argue about values, but they illustrate ways for getting beyond the stalemate that results from the mere assertion of opposing value claims. They are what the ancients called *topoi,* places in the mind where one might find arguments. Since each pattern can be used on both sides of a dispute, as the examples indicate, invoking a pattern does not by itself resolve the dispute either. Rather, it opens a space for argument, in which each of the disputants attempts to convince a relevant audience that his or her value should be preferred over that of the antagonist.

Tactics of Argument

In the ensuing discussion, the range of supporting arguments is potentially without limit. Two types, however, loom especially large. One is the role of analogy. In attempting to show that one's own value should be favored, arguers often try to show that the situation they are discussing is basically like one in which the value unquestionably applies. The logic of the argument is like this: value A clearly prevails in situation X (chosen because it is a paradigm case); this situation is basically like situation X; therefore, value A applies in this situation as well. The power of the analogy is that it uses a known and clear case to frame our understanding of a difficult or ambiguous case. If we see a strong resemblance between the two cases, then the rule of justice (Perelman & Olbrechts-Tyteca 1958/1969) dictates that we treat them both in the same way, by applying to the case at hand the same value that governs the paradigm case. The antebellum slavery debate in the U.S. illustrates the point. The status of the slave was ambiguous: was it more like that of a human being or more like that of property? Pro-slavery advocates often maintained that a slave was basically like any other class of property and should be treated accordingly, whereas anti-slavery advocates held that the slave was more like a person and therefore was entitled to personal liberty. All manner of examples from history, from other cultures, and from the Bible were used to support each analogy. Eventually, as we know, the view prevailed that the humanity of the slave trumped the status of slave as property, and once that happened, slavery became morally unacceptable.

A second type of supporting argument frequently used in these value disputes is the circumstantial *ad hominem*. In popular usage, *ad hominem* is often described as an unwarranted personal attack that diverts from the substance of the argument. As Walton (1998) has demonstrated, however, there are several different types of arguments against the person, not all of which are fallacious. A particularly potent argument is the circumstantial *ad hominem*, which claims that a person's own behavior (or circumstance) is at odds with the value he or she espouses. The implication is that the person does not "really" hold the value and therefore that it should not carry great weight. The classic example is the chain smoker who admonishes his child not to smoke and who is met with the retort, "You can't really mean that; after all, you smoke three packs a day." There are answers to this retort, of course, such as pointing out the debilitating effects of addiction, but on its face the retort suggests that the parent does not practice what he preaches and that, for that reason, the preaching should not be taken seriously. A recent example of the circumstantial *ad hominem* involved the Supreme Court's decision in *Bush v. Gore*, which essentially settled the 2000 Presidential election (Zarefsky 2003). Because the decision was an exercise in judicial activism by a Court which renounced judicial activism, critics alleged that it was not a principled or sincere decision but a political intervention by the Justices to assure the election of the candidate they had favored.

The reason circumstantial *ad hominem* plays a large role in value disputes is that objecting to a value, pointing out its limitations, or asserting a counter-value is often not enough to defeat the value. Values reflect world-views and the objections often presuppose a different world-view, one that the original advocate would simply dismiss. Another pre-Civil War example will illustrate the point. When Abraham Lincoln said he was opposed to slavery because it was wrong, Stephen Douglas replied that it did not matter that Lincoln thought slavery wrong; the people competent to decide that question were those who actually were going to the new territories to live there. When Douglas defended this version of "popular sovereignty," Lincoln answered that it made sense only if one did not believe slavery to be wrong, because one could not maintain coherently that a person had a right to do what was wrong. Lincoln's and Douglas's values grew out of incommensurable world-views, so each could dismiss the other's position as irrelevant (Zarefsky 1990).

In contrast, the circumstantial *ad hominem* holds that a value is not acceptable *to the person who expresses it,* because that person's actual behavior undercuts the value. Other things being equal, this realization deprives the arguer of the ability to espouse the value. For example, it is perfectly appropriate to decry prostitution, but it was not possible for former New York Governor Eliot Spitzer to do so after he was exposed as the client of a prostitute. Many people could declaim against extramarital affairs, but it was difficult for Spitzer's successor, Governor David Paterson, to do so after acknowledging that he had had affairs. And while many public officials could insist that citizens have an obligation to pay the taxes they owe, it was hard for Treasury Secretary Timothy Geithner to say so once it was revealed that he had owed taxes that were not paid until shortly before his nomination was announced.

Perhaps realizing the power of the circumstantial *ad hominem,* Lincoln and Douglas employed it freely in their famous debates. Lincoln held that Douglas did not really support popular sovereignty, since he had opposed an amendment to the Kansas-Nebraska Act that would have explicitly allowed Kansans to reject slavery via a public referendum. Douglas countered that Lincoln was not really willing to tolerate slavery where it already existed (as Lincoln repeatedly had insisted he would do); since he said that the country must become "all one thing or all the other," he must really be an extreme abolitionist (Zarefsky 1990).

What these six general patterns (strategies) and two specific types of support (tactics) suggest is that, notwithstanding the difficulty, people do in fact argue about values and can do so productively. It follows, then, that the moral issues posed by bioethics should not be regarded as beyond the pale of public moral argument.

THE INCONCLUSIVENESS OF MORAL ARGUMENT

From the examples, it is evident that many arguments about values are not conclusive. The very same sorts of warrants are available to advocates on any side of a dispute, and their task is to gain others' agreement that their value best fits the case. The outcome is unknown and may not be the same in each case.

In these explanations, democracy has been privileged over morality. Even the prophet, who claims to know *for sure,* ultimately must make a case that will be acceptable by others. This point of view may be readily accepted by secularists who see that the alternative is tyranny. It may be readily accepted by those whose faith traditions, like mine, hold that prophecy ceased many centuries ago, with the prophet Malachi being the last. On this view, the word of God is found not in the human voice but in sacred texts that we must struggle, with all our imperfections, to interpret. And since it is not given to us to know what they mean, *for sure,* we must recognize and respect the views of others as well as our own. This is nowhere better captured than in the Talmud, which is argumentative through and through. Human beings must decide, they cannot know for sure, so they must submit their claims and reasons to the judgment of their fellows.

But this resolution will not sit well with those of other faith traditions who believe that God continues to speak directly to us, telling us how to behave in the world. Suppose that we really did know, *for sure*, when human life begins, or what our responsibility to the planet is, or whether a particular war is a moral obligation. If we really knew *for sure*, would we be tolerant of ignorant people who did not see the light but who nevertheless challenged our judgment? Would we spare any effort to be sure that we prevailed? Would we be patient with the niceties of democratic decision-making? If we knew the truth and others decided in error, would we not also be implicated in the sin of the whole?

This is the position of the prophet. Others might call him or her a fanatic, deluded perhaps, most likely presumptuous and arrogant, but most prophets have been similarly reviled. How can we maintain that even one who claims to know *for sure* should be constrained by the proceduralism of democratic society? The answer might be to observe that it is democracy that creates the conditions in which one can espouse what he or she claims to know *for sure*. Otherwise, the tyranny of the ignorant could silence the true believer, confining him or her to ineffectual martyrdom. How religious freedom came to the U.S. is an interesting case in point. It did not grow out of Enlightenment political philosophy so much as out of practical circumstance. The Great Awakening of the mid-18th century led to a multiplication of religious sects, most of which thought that they knew *for sure*. But with greater numbers of sects, there was greater risk that any sect would be in the minority, subject to persecution—in the absence of some concept of religious freedom. So the norm developed: religious denominations must eschew force, winning adherents by argument instead. In return for accepting that tenet of democracy, each sect is free to make its case and appeal for believers. Isn't this preferable to a holy war in which one might be on the losing side?

The answer to the prophet's claim to override democracy can be made in the form of a circumstantial *ad hominem*, maintaining that a person's claims are inconsistent with his or her own circumstances. Consider the case of North Carolina's distinguished Senator Sam Ervin, who in the early 1970s chaired the Senate committee investigating Watergate. Ervin related the experience of a late-night caller from Kentucky who told the Senator that he had personal revelations from the Lord and asked that he be called as the Watergate committee's first witness, as the Almighty Lord instructed. Ervin "advised him I hated to disobey the Almighty's instruction, and we'd be delighted to welcome the Almighty as the lead-off witness, but we couldn't permit the informant to enact the role because he didn't know anything about Watergate except what the Almighty had told him and somebody might object to his testimony because it was hearsay" (Ervin 1980; Zarefsky 1987). If the caller had direct access to God, as he claimed, then surely he could invite God to testify before Ervin's committee. His inability to do so called into question whether he really had direct access to the Almighty. (Of course, in this case Ervin's Kentucky caller would not find this funny. He knew, *for sure*, that God had spoken to him.

If he was reviled and scorned, he was in good company; so too were the prophets of old.)

Even in ancient times, however, there were true prophets and false prophets, and only in the fullness of time could people know which was which. Individuals hear the call of conscience in different voices, and they trace it to different sources. But in a democratic society, moral authority comes from the ability to make arguments, grounded both in moral principle and in the circumstances of a specific case, and to gain the assent of one's fellows. The tension between democracy and morality is thereby both persistent and productive.

REFERENCES

Booth, W. C. (1974). *Modern dogma and the rhetoric of assent.* Notre Dame, IN: University of Notre Dame Press.

Douglas, S. A. (1961). Letter to twenty-five Chicago clergymen. In R.L. Johannsen (Ed.), *The letters of Stephen A. Douglas.* Urbana: University of Illinois Press. 300–322. (Originally written in 1854.)

Ervin, S. (1980). *The whole truth: The Watergate controversy.* New York: Random House.

Lincoln, A. (1953a). "A house divided": Speech at Springfield, Illinois. In R. L. Basler (Ed.), *The collected works of Abraham Lincoln.* New Brunswick: Rutgers University Press. 2, 461–469. (Originally delivered in 1858.)

Lincoln, A. (1953b). Message to Congress in special session. In R. L. Basler (Ed.), *The collected works of Abraham Lincoln.* New Brunswick: Rutgers University Press. 4, 421–441. (Originally delivered in 1861.)

Perelman, C. & Olbrechts-Tyteca, L. (1969). *The new rhetoric: A treatise on argumentation.* Trans. J. Wilkinson and P. Weaver. Notre Dame, IN: University of Notre Dame Press. (Originally published in French in 1958.)

Schulman, A. (2008). Bioethics and the question of human dignity. In *Human dignity and bioethics: Essays commissioned by the President's council on bioethics.* Washington: U.S. Government Printing Office. 3–18.

Thorson, T. L. (1962). *The logic of democracy.* New York: Holt, Rinehart, and Winston.

Walton, D. (1998). *Ad hominem arguments.* Tuscaloosa: University of Alabama Press.

Zarefsky, D. (1987). Fulbright and Ervin: Southern Senators with national appeal. In C. M. Logue and H. Dorgan (Eds.), *A new diversity in contemporary Southern rhetoric.* Baton Rouge: Louisiana State University Press. 114–165.

Zarefsky, D. (1990). *Lincoln, Douglas, and slavery: In the crucible of public debate.* Chicago: University of Chicago Press.

Zarefsky, D. (2003). Felicity conditions for the circumstantial ad hominem: The case of Bush v. Gore. In F. H. van Eemeren, J. A. Blair, C. A. Willard, & A. F. Snoeck Henkemans (Eds.), *Proceedings of the Fifth Conference of the International Society for the Study of Argumentation.* Amsterdam: Sic Sat. 1109–1114.

Zarefsky, D. (2008). Two faces of democratic rhetoric. In T. F. McDorman and D. M. Timmerman (Eds.), *Rhetoric and democracy: Pedagogical and political practices.* East Lansing: Michigan State University Press. 115–137.

2 Bioethical Deliberation in a Democracy

Jonathan D. Moreno

Professor Zarefsky and I couldn't have choreographed our essays better. He has discussed a number of themes that I am also going to talk about in the context of the field of bioethics. I recognize that many readers don't eat, sleep, and drink bioethics, so I am going to try and do some suitable translating. What I am talking about should be fairly clear, because so many of the themes I'll mention are familiar to readers just by virtue of their being informed citizens. My focus here is what I like to call the moral paradox of consensus. My argument concerning this paradox initially took form in my book *Deciding Together: Bioethics and Moral Consensus* (Moreno 1995). As many know, this became a Broadway play and bestseller and I was played by Ben Affleck in the movie, but it was his first role, so it didn't get a lot of attention.

The moral paradox of consensus has three parts. Part one entails the notion of moral heroism that comes to us from the Western philosophical tradition, and that is first and foremost associated with the life and death of Socrates. There are many visual images of Socrates about to drink the hemlock. Art historians may have a lot to say about these images and the iconography of heroism they represent. I was not in the room at the time—I was only a graduate student—but the point of the imagery is clear: Socrates is the great man, who has insight into the good and whose death is an occasion for shame and regret. The several accounts of the death of Socrates are remarkably consistent, and very moving. In essence, Socrates was a martyr for truth and goodness. He drank the hemlock rather than accept the opportunity to bribe the guards and leave Athens. He had been prosecuted on the grounds that he had corrupted the youth of Athens and "made the worse the better cause." In other words, he was deemed a lawyer, a sophist, leading youth to doubt the wisdom of state leaders. The image of Socrates, this good man, is the paradigmatic image we have of moral heroism, along with others, like Moses, Jesus, St. Joan and Dr. King.

Moral heroism is one of the most powerful ideas we have in the Western tradition. It speaks of the possibility that individuals can exhibit the character, commitment, and courage to do what is right regardless of the consequences to themselves. Indeed, heroes "stand out" from the crowd, and are (eventually) admired for doing so. Heroes help to direct a culture's moral compass by offering models of human greatness. Heroism sounds a call of conscience, a call that incites moral consciousness and beckons us to be upstanding and outstanding citizens. The corollary of moral heroism, however, is that the mob, the *demos*, is untrustworthy. Not only is the mob likely to reach the wrong conclusion; it may be impossible for the group, the mob, to reach the right conclusion, simply by the nature of mob psychology; hence, the moral paradox of consensus. There are other interpretations of the story of Socrates, of course, but I think that this is the prevailing one.

Moving now from ethics to bioethics means talking about values in medical science and in modern institutions. As Carl Elliott's essay suggests, bioethics is often thought to have emerged from the 1960s,[1] especially in the U.S., as part of the same set of concerns that gave rise to the civil rights movements of that era. The book and film *One Flew over the Cuckoo's Nest* (Kesey 1963; Forman 1975) exemplified the struggle of psychiatric patients who demanded their civil rights, their civil liberties. In many ways, modern bioethics could be considered to be a kind of protest movement, a social reform movement in the context of great institutions, great complicated bureaucracies like hospitals.

As Max Weber (1947) and many other sociologists have pointed out, modern bureaucratic institutions cannot function without consensus. Think of all the things that have to happen for staff, students, and faculty to come to the university every morning, and for that university to operate. Think of all the different people who have to engage in different procedures, do all kinds of different jobs. Without coordinating their activities, they just do them. These are institutions that presuppose consensus. They presume that people have to come in and do their jobs and that it's okay to work for a university. The philosopher's myth of the social contract bases consensus on active agreement. Here, "myth" is not a pejorative term; rather, it's a story that certain philosophers tell to explain consensus. However, unlike that social contract mythology, at universities we don't actually ever sit together and say, "Tomorrow morning you people who work as administrators in the economics department are going to come in and do such and such, and then I as the professor am going to come in (maybe I will come in and maybe I won't . . . I am the professor) and treat my students a certain way, and when you, students, come into my classroom you agree that you are going to sit quietly and appear to be listening to the speaker." One of the reasons this bureaucracy called the university, or the academic world, is so successful is that we at universities are so good at acquiescing. The reality is that we are all very well-socialized. We know what the rules are. We are very cooperative and we engage in a kind of passive consensus. This consensus can be fully understood and appreciated only in the historical context of preconditions and preconceptions that define our everyday

existence and its "worlds of know-how": those domains of traditions, cus-
toms, rules, norms, and routines that inform the common and "normal"
ways we perceive, think, and act in our daily lives. The everyday world of
know-how provides an abundance of habitats or environmental settings for
living in accordance with societal norms and standards. This is part two of
the paradox: we can become so good at much of what we do only because
we do it without giving it a second thought. Passive consensus is not agreed-
upon; it is "automatic pilot."

Returning to bioethics, the paradox, part three, is that bioethics arose
in a culture, particularly in the U.S. in the later 1960s, steeped in images
of idealism and moral heroism. Carl Elliott (this volume) and I (Moreno
2001, pp. 240–243) have both written about Professor Henry Beecher at
Harvard, who wrote a famous 1966 paper in the *New England Journal
of Medicine*, in which he reported on a couple of dozen cases of unethical
use of human subjects in the published literature and medical experiments
(Beecher 1966). The interesting thing about Beecher is that he was not born
a Beecher; he was born Henry Unangst in Cape Girardeau, Missouri, and
when he went to medical school in Boston he changed his name to Beecher.
You might wonder: why Beecher in Boston? Names like Harriet Beecher
Stowe and Henry Ward Beecher both signal the great tradition of aboli-
tionism. It is better to be a Beecher in blue blood Boston than it is to be an
Unangst if you are an ambitious young man. And it happened that on his
mother's side, Henry was related to the Beechers.

So Beecher became a kind of abolitionist. He became a spokesman, a
moral hero with respect to the civil rights of people in human experiments.
I think he actually saw himself that way; he was part of that 1960s mindset.
(His previous career as an LSD experimenter and consultant to the CIA is
another story (Moreno 2001, 2006).) Beecher saw that there was a need to
make a positive statement about the rights of human beings in the context
of research experiments. Moreover, he did not believe that the way to solve
the problem of exploiting human beings in experiments was through rules.
He was not in favor of bureaucratic mechanisms like institutional review
boards (IRBs), or—God forbid—bioethicists. He believed in the virtue of
the individual experimenter. This returns us to that old Socratic/Platonic
tradition of the individual who knows the good. Not the community; not the
university; certainly not the Food and Drug Administration. This is a really
interesting paradox: we have a field that emerged from a concern about
individual rights, modeled on certain kinds of moral heroism by individuals
able to see the right thing to do; but how do we advance the goals of that
field? Through committees. It is a very odd situation in which we do ethics
through committees. We are so inured to the idea of ethics committees that
we don't even see the irony anymore. Bioethics arose in a culture steeped in
the imagery and idealism of moral heroism, but responded to the demands
of a bureaucratized health care system, which leaves it torn between the
values of individual autonomy and the values of institutional solidarity. This

requires us to ask whether ethics committees are sources of "group think" or avenues for the earnest pursuit of (approximate) moral truth.

When I started working in bioethics in hospitals, many senior doctors would say, "You know, Dr. Moreno, if I need a committee to tell me how to do the right thing I am going to leave medicine." That is the sensibility: that the individual virtuous doctor knows the good. S/He doesn't need a committee to tell him or her. That has interestingly somewhat changed. Maybe it is a generational change. But now I think we have gone to the other extreme. I have heard doctors say "I can't really make a moral judgment. I am going to send it to the ethics committee." This is where we are now, I think, with respect to the moral paradox of consensus in bioethics committees.

Now I want to discuss the role of public commissions in bioethics. Right now in this country we are going through a highly relevant debate about whether we ought to have a public commission on the interrogation and detention policy employed post-9/11. There is some reason for people in bioethics to be interested in this; after all, some of the people who have participated in the interrogation exercises currently being questioned were psychologists, and there are interesting medical ethics questions that arise about the role of the psychologist in doing these interrogations or in helping to prepare the conditions for these interrogations. The problem is that the very decision to establish a public commission of some sort that involves ethical questions itself presupposes a series of other commitments, and those are commitments that President Obama has said he really isn't willing to make. One of those commitments is whether we should spend a lot of time and public energy on this question, whether it is a priority at this point in our history. I think it is interesting that we are going through this commission question now. The other problem with commissions is that, as people who have worked for commissions or who have been on them know, it is possible to stack the deck by deciding who will be on a commission and who will not. Is it going to be a commission of members of Congress or is it going to be a commission of citizens? How are the members selected? This turns out to be extremely complicated.

Interestingly, in the late 1960s, the U.S. Senate considered creating the first of what we would today call bioethics commissions, years before the U.S. Public Health Service syphilis study was well-known. It came out of the same spirit of the age, the same zeitgeist, as the Hastings Center, the first bioethics think tank, which is now celebrating its 40th anniversary. The so-called "Georgetown mantra," the familiar principles of bioethics—autonomy, beneficence, nonmaleficence, and justice—first surfaces in a national bioethics commission report in the late 1970s following revelations about the syphilis study (National Commission 1979). Ironically (and this is part of a longer story), the greatest of what I call fissures in the field of bioethics, that is, the greatest personal differences in the field, arose in the last few years around the President's Council on Bioethics under its first chairman, Dr. Leon Kass. I have called this crisis in institutional bioethics the end of

the great bioethics compromise (Moreno 2005). The idea of the great bioethics compromise was basically this: from the beginning of the field in the
late 1960s through the late 1990s, about 30 years, although there were differences among bioethicists, for example, about the beginning of life, these
differences were not, for the most part, allowed to pervade the field. People
said, "Well, we disagree about some of these basic questions about the
origin of life, but we have a lot of other work to do, so we are not going to
talk about that a lot." In fact, there were very few people in bioethics who
talked, as professional bioethicists, about the abortion issue. That began to
change; that great bioethics compromise began to fracture about 10 years
ago, first with the birth of Dolly the cloned sheep, and then with the stem
cell debate. There have been, I think, greater political divisions in public
among bioethicists as a result of these developments. The great bioethics
compromise has been stressed, at best; perhaps fractured.

This history provides us with an opportunity to think about how we
arrive at consensus. Professor Zarefsky describes very well a rhetorical
and moral technique that some bioethicists and philosophers have recommended: casuistry. The basic idea of casuistry is to identify a line of "cases
of conscience" (*casus conscientiae*) and compare them. Casuistry begins
from what is called a paradigm case, which is supposed to be the exemplary
case for the fulfillment of a certain moral principle. Other cases are then
compared to the paradigm case. Through analogy and contrast and comparison—which may be done individually or in a group, for example as a
jury, a committee, or commission—we come out with a certain view of how
far this case is from the paradigm case. In many ways this is a very pragmatic approach, because it is meant to be case oriented rather than driven
by *a priori* theory. We can agree upon a result, even while we acknowledge
that there are many ways of arriving at that result. In their classic treatment
of the topic, the philosophers Albert Jonsen and Stephen Toulmin (1988)
note that the goal of casuistry is "to set out descriptions of moral behavior
in which moral precepts and the details of action [are] looked at together"
in order to determine how people, "concerned to act rightly," should "make
a judgment of conscience in a specific kind of situation." Cases of conscience require us to be rhetorically competent in order to engage others in
collaborative deliberation about the meaning of the moral precepts found
in case examples and their relationship to everyday experience.

Casuistry, as Professor Zarefsky's essay demonstrates, is very much in the
spirit of American pragmatism. Focusing on cases rather than on abstract
principles fits well with the way 20th century American moral culture developed. However, casuistry's relationship to reasoning from moral principles,
as exemplified in the mantra (autonomy, beneficence, nonmaleficence, and
justice), was somewhat unclear at the beginnings of bioethics scholarship.
As a result, when I wrote *Deciding Together,* one of my reviewers said
that I talked about so many different kinds of consensus, and so many
distinctions, that he was led into an alpha state—which I thought was not

a compliment. I will not describe here all the different senses of consensus that I wrote about in my philosophy mode, but I do want to distinguish between a few of them.

First of all, there is superficial consensus and there is deep consensus (Moreno 1995). Superficial consensus is generally all we require when we are in a committee or a jury. In other words, do we agree on what ought to happen in this case? Do we agree on the punishment? Do we agree on the remedy? Do we agree on the facts? That's a superficial consensus (ibid.). Now when you start digging deeper, you might find out that people got to that superficial consensus in very different ways, so that deeply they are not in consensus. Now, that could make a big difference to you. You might say, "Gee, you know, when we all thought we agreed, I was fine. Now I am a little worried, because your way of getting there was so different from mine; there must be something wrong here." The ironic conclusion that suggests itself here is that when you are on a jury it may be better not to know too much about the underlying moral values of the other people you have to agree with.

But closer examination of a superficial consensus can also be revealing and, at least occasionally, essential. The point is dramatically made in Sidney Lumet's famous film *12 Angry Men* (1957). The film tells the story of members of a jury in a murder trial who after only 10 minutes or so reach an 11–1 decision that the accused young man is "obviously" guilty of killing his father. Owing to the lone dissenter's courage, commitment to detail, rational judgment, and rhetorical competence, however, the vote begins to reverse itself as the other jury members are made to see how their prejudiced and stereotypical judgments are incommensurate with the open-minded and open-hearted collaborative deliberation needed to disclose the truth of the matter at hand. Eventually, all 12 men conclude unanimously in favor of an acquittal, not because they think the boy is innocent but because each juror has come to a "reasonable doubt" about his guilt. Construction of a deep consensus is thus always a struggle, but it can be educational and life-saving, a source of wisdom and compassion.

Yet even superficial consensus is more than mere compromise. The philosopher Martin Benjamin wrote a wonderful book about compromise: a process where you are basically trying to preserve your principles by, as he says, "splitting the difference" (Benjamin 1990). The moral and rhetorical process of consensus, however, requires more than this, for with this process you are trying to find the *right and just* conclusion together as a group.

Finally, it is important to distinguish between consensus as product and process. I consider this distinction extremely important. Consensus is a noun and that's too bad. What we really do is "consensize." It's a verb. Nouns give us the unfortunate impression that there is a something out there, a state to get to, as in "reaching" consensus. Lincoln didn't reach consensus. Lincoln struggled. And the nation struggled. It consensized at the cost of 600,000 lives. I have just read the new Harvard President Drew

Gilpin Faust's powerful book, *This Republic of Suffering: Death and the American Civil War* (2008). It gives reaching consensus a whole new meaning to read about how people died during the Civil War. It is not an abstract process, and it is often not a pretty process; but it is a process.

How do we justify consensus? Actually, a thoroughgoing Platonist cannot justify consensus. It's not justifiable. What is wanted by Platonists is instead a "natural aristocracy" of people who, like Socrates, are devoted to knowing the good. According to social contract theory, consensus conclusions can be justified when they are reached under Rawls' veil of ignorance (1971, pp. 136–142). The relevant principles are deemed rational *a priori*. According to pragmatic naturalism, the view that I happen to hold, conclusions may be justified on the basis of the process used to reach them. The trick here, the difficult and disconcerting part, is that this approach to moral justification may be very unsatisfactory to people who prefer to be sure that they are always right. Whatever consensus you reach, according to this third approach, must always be revisable. In other words, it has to be, as the philosophers of science would say, "falsifiable." Or as 19th century philosopher Charles S. Peirce says, we must see ourselves as fallible (Margolis 1998). Peirce too came out of the Civil War generation. That generation knew what falliblism was about, understood the horror of that war. Or, as the late philosopher Richard Rorty said, we need to be "liberal ironists" (Rorty 1989). This is the hard part. We need to believe very deeply in our values, and at the same time we need to know that we could be wrong. That is a very difficult dance to do. It exposes the self to uncertainty and to the anxiety engendered by this highly dynamic state of existence. Yet this dance is central to the production of knowledge in the modern world. Consider the words of Nobel Prize winning physicist Richard Feynman: "Every scientific law, every scientific principle, every statement of the results of an observation is some kind of a summary which leaves out details, because nothing can be stated precisely." Feynman thus maintains:

> All scientific knowledge is uncertain. This experience with doubt and uncertainty is important. I believe that it is of very great value, and one that extends beyond science. I believe that to solve any problem that has never been solved before, you have to leave the door to the unknown ajar. You have to permit the possibility that you do not have it exactly right. Otherwise, if you have made up your mind already, you might not solve it . . . Scientists are used to this. We know that it is consistent to be able to live and not know. Some people say, "How can you *live* without knowing?" I do not know what they mean. I always live without knowledge. That is easy. How you get to know is what I want to know (Feynman 1998, pp. 25–28).

Feynman explains that this "freedom to doubt" allows science to thrive, and describes his felt responsibility as a "citizen-scientist" to "proclaim the value of this freedom and to teach that doubt is not to be feared, but that it is to be welcomed as the possibility of a new potential for human beings. If you know that you are not sure, you have a chance to improve the situation. I want to demand this freedom for future generations." (Feynman 1998, p. 28)

What, then, are some of the values that support fallibilism as essential to moral consensus? Here I borrow from my friend and colleague, Tris Engelhardt. In assessing moral consensus, we first of all commit ourselves to non-violence as a means of conflict resolution (Engelhardt 1986). The Civil War was not the best way to do things. We need to be open to evidence. We need to be willing to respect unpopular points of view, and we even need to be able to ask people who have not been heard to speak—particularly in groups where they may be reluctant to speak. As the philosopher John Dewey, the pragmatic naturalist, held, the conclusions we reach must be revisable (1941). In parallel with Feynman's view of the way to acquire scientific knowledge, this is a kind of experimentalist view of the way we approach consensus.

Now let me go back to the problem of representation, which I raised in connection with the question of membership on commissions. In ethics, when government decides to create a commission, there are basically three kinds of commissions to create. This problem will arise if the senate decides to establish a commission on the interrogation matter. First, a commission may be built around stakeholders: that is, people who have a particular stake in, a particular interest in, the issue, which will presumably include people who have been or feel aggrieved. If we wanted to (yet again) reform the system for conducting clinical trials, we would have people on that commission who represent people who are in clinical trials or people who are from disease advocacy groups as well as those who actually conduct clinical trials. A second kind of commission is one composed of experts; thus, ethical experts would be appointed to a bioethics commission. Whether there is ethical expertise is an interesting question. Most of the commissions that are created by government are expert commissions. For example, if the Senate decides to create a commission to investigate the detention and interrogation policies and practices over the last eight years, they will appoint a whole lot of lawyers. I would expect to see somebody like Watergate lawyer Richard Ben-Veniste or somebody very much like him on that commission.

Government commissions tend to be made up of the people who know how to operate in that kind of environment; thus, they are often quite similar. But commissions could also be made up of people who share the same philosophical presuppositions. A commission established to look at a certain controversial issue could be composed of people who have certain sympathies up front for one view or the other with respect to that issue. Of course these three varieties of commission are not pure types; they can be mixed. So I raise the question—what counts as a bioethics commission? We could have an interesting discussion about what bioethics itself is, let alone what a bioethics commission is, but I am going to use the folk language.

A few examples of government bioethics advisory committees. Fifteen years ago, I served on the staff of a presidential commission on human radiation experiments.[2] That was an amazingly interesting project. There were rumors for decades of unethical federally sponsored experiments involving human subjects with ionizing radiation, dating back to the Second World War. There were allegations that these experiments were covered up, and that many people were injured as a result. This advisory committee is not often thought of as a bioethics commission, perhaps because the first half of what we did between 1994 and 1995 was more like a truth commission, more like what people have in mind for the interrogation problem. Because we had a presidential mandate, we were able to go into classified files and actually make findings of fact at the end of the year and a half. We also reached moral conclusions about what was done.

The National Human Research Protection Advisory Committee (NHRPAC), was one of dozens of standing advisory committees within the Department of Health and Human Services (DHHS). Between 2002 and 2008, NHRPAC gave guidance on the federal regulations governing research involving human beings, also known as the Common Rule. Clearly NHRPAC was (and its current successor committee, the Secretary's Advisory Committee on Human Research Protections, is) in some sense a bioethics committee. Many other committees within the federal government also give bioethics advice even though that is not their explicit charge, for they weigh risks and benefits, worry about informed consent and so forth. Examples are FDA drug approval advisory committees.

I served as co-chair of another committee, the National Academy of Sciences/Institutes of Medicine Committee on Guidelines for Human Embryonic Stem Cell Research, addressing guidelines for human embryonic and stem cell research. What is different about this committee from the first two is that the National Academy of Sciences is not a government agency, so unlike the first two entities, we were not required to have our meetings in public. There is a federal law, the Federal Advisory Committee Act, which requires that any commission created using your tax dollars to advise the president, the cabinet secretary, or other parts of the government must hold all of their meetings and activities in public. The National Academy of Sciences was chartered by the Lincoln administration to give advice to the federal government, but doesn't have to have its meetings in public. Committees of the National Academies are quasi-public expert commissions.

Finally, I want to turn to the controversy around the President's Council on Bioethics in the first Bush term. The President's Council's original agenda was focused on the issues of human dignity; the role of biotechnology in our lives and our society; and what it means to be human. Recall that it was created and its membership appointed in the wake of controversies over embryonic stem cells and cloning. It produced reports on human dignity, enhancement, and similar issues.[3] Although there was much criticism of the intellectual orientation of the most influential members of the Council, what I found most interesting was the fact that its charter explicitly excluded consensus as a goal of the Council's deliberations. Evidently those who participated in developing the Council's charter believed that consensus too often suppresses important viewpoints, that it thereby invites slippery slope reasoning. Whether that was a practical standard or even whether the Council finally functioned in accord with it are not questions I can pursue here. I only note that the formal anti-consensus orientation distinguished it from other bioethics commissions. At this writing, the appointment of President Obama's own bioethics council was underway. Based on the published charter of the Obama commission we can conclude that it will not adopt an anti-consensus stance.

The philosopher Jurgen Habermas said that there was a legitimation crisis in Western capitalism (1975). (That was before the fall of the Soviet Union. Interestingly enough, Western capitalism is having another legitimation crisis of its own doing now.) Is there a legitimation crisis in bioethics? I think there is. I am not sure that we need to lose any sleep over it, but there is an interesting problem in the field of bioethics, which is significant for the pursuit of public deliberation about moral questions that relate to bioethics issues. There are underlying tensions in the field: between the left and the right; between pro-life conservatives and pro-choice progressives; between libertarian conservatives and neoconservatives, and between progressives who are more and those who are less enthusiastic about technological innovation. These and other underlying tensions have been largely unresolved since the 1970s, and they have now become very visible. This poses a challenge for people in bioethics, who like to think of themselves as able to help to craft moral consensus with and for the public. It raises a very difficult question: is bioethics merely ideology masquerading as public moral philosophy? Tristam Engelhardt has answered this question in the affirmative (Engelhardt 2002).

I think the answer is no; but it is a question that we need to ask ourselves. The best answer will be framed in terms of the democratic process values that I have described, so that we can achieve moral consensus without the eruption of another Civil War.

NOTES

1. Whether one believes "bioethics" is a field qualitatively different from traditional medical ethics depends partly upon the significance one attributes to the emphasis on medical truth-telling and patient autonomy in the discussions that began in the late 1960s—which differed from the more global concept of bioethics first introduced by Van Resselaer Potter (1971). More speculative topics, like the meaning of the decoding of DNA for society, were also not part of traditional, doctor-oriented medical ethics.

2. The Advisory Committee for Human Radiation Experiments, 1994–95. Its report is available at http://www.bioethics.gov/commissions/.
3. Reports from the President's Council (as well as from NHRPAC and SACHRP) may also be found at http://www.bioethics.gov/commissions/.

REFERENCES

12 Angry Men (1957). Directed by Sidney Lumet. MGM.

Beecher, H. K. (1966). Ethics and clinical research. *New England Journal of Medicine* 274: 367–372.

Benjamin, M. (1990). *Splitting the difference: Compromise and integrity in ethics and politics.* Lawrence, KS: University Press of Kansas.

Dewey, J. (1941). Propositions, warrented assertibility, and truth. *Journal of Philosophy* 38(7): 169–186.

Engelhardt, H. T. Jr. (1986). *The foundations of bioethics.* New York: Oxford University Press.

Engelhardt, H. T. Jr. (2002). Consensus formation: The creation of an ideology. *Cambridge Quarterly of Healthcare Ethics* 11(1): 7–17.

Faust, D. G. . (2008). *This republic of suffering: Death and the American Civil War.* New York: Alfred A Knopf.

Feynman, R. P. (1998). *The meaning of it all: Thoughts of a citizen-scientist.* Reading, MA: Perseus Books.

Habermas, J. (1995). *Legitimation crisis.* Translated by T. McCarthy. Boston: Beacon Press.

Jonsen, A. R. & Toulmin, S. (1988). *The abuse of casuistry: A history of moral reasoning.* Berkeley, CA: University of California Press.

Kesey, K. (1963). *One flew over the cuckoo's nest.*

Margolis, J. (1998). Peirce's fallibilism. *Transactions of the Charles S. Peirce Society* 34(3): 535–569.

Moreno, J. D. (1995). *Deciding together: Bioethics and moral consensus.* Oxford: Oxford University Press.

Moreno, J. D. (2006). *Mind wars: Brain research and national defense.* New York: Dana Press.

Moreno, J. D. (2001). *Undue risk: Secret state experiments on humans.* New York: Routledge.

Moreno, J. D. (2005). The end of the great bioethics compromise. *Hastings Center Report* 35(1): 14–15.

National Commission for Protection of Human Subjects of Biomedical and Behavioral Research (1979). *The Belmont report: Ethical principles and guidelines for protection of human subjects of biomedical and behavioral research.* Available from <http://odsr.od.nih.gov/guidelines/belmont.html> Accessed 7 June 2010.

One Flew over the Cuckoo's Nest. (1975) Directed by Milos Forman. United Artists.

Potter, V. R. (1971). *Bioethics: A bridge to the future.* Upper Saddle River, NJ: Prentice Hall.

Rawls, J. (1971). *A theory of justice.* Cambridge, MA: Belknap Press of Harvard University Press.

Rorty, R. (1989). *Contingency, irony, and solidarity.* Cambridge: Cambridge University Press.

Weber, M. (1947). *The theory of social and economic organization.* Trans. A. M. Henderson and T. Parsons. London: Collier Macmillan Publishers.

3 Bioethics and the Law
Using Moot Court as a Tool to Teach Effective Argumentation Skills

Christine Nero Coughlin,
Tracey Banks Coan, and Barbara Lentz[1]

Effective argumentation is essential for resolving bioethics issues. This chapter examines how moot court, a tool used in legal education, can bolster students' capacity to effectively argue in a bioethics context. This chapter first sets forth a sample moot court problem and corresponding oral argument text that can be used to teach a bioethics issue. The chapter then identifies challenges to teaching effective argumentation in bioethics and explains how and why moot court augments students' capacity to identify competing ethical and moral visions, and to objectively evaluate various viewpoints in bioethics decision-making. The chapter concludes with identifying some possible limitations to the use of moot court in academic settings other than law school, and provides solutions to overcome the limitations.

SAMPLE MOOT COURT CASE FILE

The following sample moot court case file was successfully used in a bioethics course last taught by Professors Mark Hall[2] and Christine Nero Coughlin at Wake Forest University School of Law in 2006. The case deals with a fictional lawsuit filed by the Does against Wake County Hospital requesting that their baby, born with anencephaly, be declared legally dead so that her organs could be donated for purposes of transplantation. The sample includes a fictional fact scenario that was distributed to all students at the beginning of the course and the text from oral arguments that Professors Hall and Coughlin, who represented the Does and Wake County Hospital, respectively, presented to a mock panel of student judges.

THE BABY JANE DOE CASE

At or about the eighth month of pregnancy, the parents of the child Baby Jane Doe were informed that she would be born with anencephaly. Anencephaly is a birth defect, invariably fatal, in which the child typically is born with only a "brain stem" but otherwise lacks a functioning brain. The first sign of trouble had come when Mrs. Doe was being examined at the Wake Family Health Clinic in the State of Wake, where she was enrolled in a prenatal program for indigent mothers. On the little television screen above her bed floated the image of her baby. As the technician ran his finger across the screen, he pointed out the liver. The heart. The stomach. The hands. The feet. The spine. The neck. And then he stopped. "Something's not right here," he told her. "What's the matter?" she asked. All he could say was, "It's the head. The top of the head." "The top of the head?" she wondered, "What on earth could that mean?"

The diagnosis, anencephaly, meant that the infant would be born with a rare, congenital disorder in which a major portion of the brain, skull, and scalp are missing. Anencephaly is both severely disabling and invariably fatal. The lack of a cerebral cortex renders the infant permanently unconscious; it is only due to the presence of a functioning brain stem that the heart and lungs operate to keep the infant alive. Such infants are incapable of any sensory perception at a conscious level, although many respond reflexively to noxious stimuli and exhibit feeding and respiratory reflexes.

The incidence of anencephaly in the U.S. is extremely low – estimated at approximately 0.3 per 1,000 births, or 1,050 per year, approximately two-thirds of which are stillbirths. In those cases where anencephalic infants are not stillborn, they rarely live more than 24 hours, and only one of seven is alive at the end of 72 hours. Although Baby Jane Doe had no hope of life herself, Mr. and Mrs. Doe opted to continue the pregnancy rather than aborting the deformed fetus because they expressly hoped that the infant's organs could be could be used for transplant in other sick children.

Baby Jane Doe was carried to term and was born alive. The back of Baby Jane Doe's skull was entirely missing and the brain stem was exposed to the air, except for medical bandaging. As an anencephalic, part of her brain stem was still functioning, but she was not conscious and would die within hours or, at the very most, days. When, however, Mr. and Mrs. Doe requested that Baby Jane Doe be declared legally dead by neurological criteria for the purpose of allowing her organs to be transplanted, her health care providers refused out of concern that they thereby might incur civil or criminal liability.

Mr. and Mrs. Doe immediately filed a petition in the trial court asking for a judicial determination that Baby Jane Doe was legally dead, as was required before her organs could be made available for transplantation. After hearing testimony and argument, the trial court denied the parents' request on grounds that the governing statute, §382.009(1) Wake Statutes, would not permit a determination of legal death until all brain activity stopped. Mr. and Mrs. Doe appealed the ruling. During the time between the trial court's decision and the scheduled appeal – nine days – Baby Jane Doe died.

The Wake Court of Appeals, however, allowed the litigation to proceed, even after Baby Jane Does' death, because the health care providers needed resolution of the question, since they were likely to face the same complaint and same short resolution timeframe in the future (Weinstein v. Bradford 1975). On appeal, the Wake Court of Appeals summarily affirmed the trial court's decision, but then certified the trial court's order to the Wake Supreme Court for immediate resolution of the issue. Assume that the Wake Supreme Court has jurisdiction and authority to rule on the merits of the case.

Wake Statutes Annotated – § 382.009. Recognition of neurological death under certain circumstances

(1) For legal and medical purposes, where respiratory and circulatory functions are maintained by artificial means of support so as to preclude a determination that these functions have ceased, the occurrence of death may be determined where there is the irreversible cessation of the functioning of the entire brain, including the brain stem, determined in accordance with this section.

(2) Determination of death pursuant to this section shall be made in accordance with currently accepted reasonable medical standards by two licensed physicians. One physician shall be the treating physician, and the other physician shall be a board-eligible or board-certified neurologist, neurosurgeon, internist, pediatrician, surgeon, or anesthesiologist.

(3) The next of kin of the patient shall be notified as soon as practicable of the procedures to determine death under this section. The medical records shall reflect such notice; if such notice has not been given, the medical records shall reflect the attempts to identify and notify the next of kin.

(4) No recovery shall be allowed nor shall criminal proceedings be instituted in any court in this state against a physician or licensed medical facility that makes a determination of death in accordance with this section or which acts in reliance thereon, if such determination is made in accordance with the accepted standard of care for such physician or facility. Except for a diagnosis of neurological death, the standard set forth in this section is not the exclusive standard for determining death or for the withdrawal of life support systems.

SAMPLE ORAL ARGUMENTS[3]

Oral Argument on Behalf of the Does

May it please the Court. My name is John Jones, and I represent petitioners Mr. and Mrs. Doe, the parents of the infant. There are two separate grounds for reversing the lower court's decision:

First, the legal definition of death should encompass anencephalic infants, who are born literally without any brain matter whatsoever other than a brain stem. Therefore, as a medical certainty, they never have and never will experience any form of consciousness. They will also cease breathing in only a short time. This makes them *sui generis*: a completely unique and self-limiting exception to the normal criteria for determining death.

Second, even if anencephalic infants are not dead, the so-called "Dead Donor Rule" should be revised to allow parents to authorize the donation of vital organs, while these infants are still alive. Both these positions are supported by all of the major principles of bioethics, namely, autonomy, beneficence, nonmaleficience, and justice.

Anencephaly is one of the most devastating birth defects known to medical science. If detected prior to birth, most parents opt for abortion, but Mr. and Mrs. Doe chose the heroic path as their way to cope with what otherwise was a tragedy: they wanted to use their daughter's organs to save the lives of other infants. The problem these parents and their physicians faced, however, was the legal uncertainty about removing organs while the infant's blood is still circulating. Unless this is done, the delicate organs usually deteriorate too rapidly to be useful. Infants are the only good source of transplantable organs for other infants. Therefore, doctors are more than willing to accept organ donations from anencephalic infants, but they are barred by the threat of legal liability. That is why we are before this court. The lives of hundreds of infants each year are at stake.

Similar circumstances confronted the President's Commission for the Study of Ethical Problems in Medicine and Biomedical and Behavioral Research in 1981. Developments in medicine created a critical shortage of life-saving organs for adults. The Commission, and certain influential state Supreme Courts, responded by revising the legal definition of death to allow retrieval of organs from patients while their hearts were still beating. The same principles underlying this universally accepted redefinition of death also support the very modest and limited evolution required in this case.

Declaring anencephalic infants legally dead is consistent with the same unitary concept of death recognized by the President's Commission, namely, "when the body's physiological system ceases to constitute an integrated whole" (President's Commission for the Study of Ethical Problems in Medicine and Biomedical and Behavioral Research 1981, p. 33). The permanent and absolute absence of any capacity for consciousness constitutes this same "collapse of psycho-physical integrity" (p. 58). Death isn't an absolute event, but a continuum. The precise point of death is socially constructed, to serve social purposes. "Legally dead" is like "legally blind" (Truog & Robinson 2003). Someone who is legally blind may still retain some sight. Someone who is legally dead according to the statutory definition of death may still retain some neurologic functioning but be reasonably treated as dead (Truog & Robinson 2003). The definitions are not precise, and for good reason.

Further, including anencephalic infants within the definition of death will not create a slippery slope or pragmatic problems for three reasons:

- The complete absence of any brain matter other than the brain stem removes uncertainty about whether consciousness is possible. Anencephalic infants have no physical capacity to develop consciousness. Those who are dead by neurological criteria have organ life. Similarly, anencephalic infants have only organ life, not life as developed persons. Their state is somewhat similar to that of an early fetus; however, whereas a fetus has potential for life as a person, an anencephalic infant does not. This is not true for others with complete physical brains who are suffering from cognitive impairment or living in a persistent vegetative state – there is potential in these patients for life beyond mere organ life.
- As with whole brain death, it is also true for anencephaly that all bodily functions will soon cease in any event. Declaring these infants legally dead before the inevitable circulation of blood has ceased and their organs become unusable does not diminish their prospects for survival. Anencephalic infants simply cannot survive.
- There are clear and irrefutable diagnostic signs of when this condition exists, thus eliminating the danger of misdiagnosis. With such a clear standard for diagnosis, infants who do not suffer from this condition will not be misdiagnosed and proposed as donor candidates.

So, allowing organ donation in this limited and specific context is an incremental, not radical, change. It is far more incremental than the original adoption of neurological criteria for death. And, like that change, courts are fully empowered to adopt this position as a matter of common law evolution. In fact, the Wake statute adopting death by neurological criteria explicitly recognizes this court's inherent authority by stating: "This section is not the exclusive standard for determining death or for the withdrawal of life support systems."

Second, even if this court chooses not to expand the definition of death by neurological criteria to cover anencephaly, it should create a limited exception to what is called the Dead Donor Rule, to allow parents like Mr. and Mrs. Doe to authorize retrieval of vital organs prior to death. Granting these parents' wishes will honor all the major principles of bioethics.

Death under these circumstances is an absolute case of nonmaleficence. Here, the good for several infants who would gain the chance of survival outweighs harm to the infant who is certain to die within a very short time. Especially in light of the parents' wishes and beliefs, this court should balance an extremely limited harm against the massive good donation of Baby Jane Doe's organs could accomplish. Several hundred infants die each year for want of transplanted organs. Several hundred anencephalic infants are also born each year. Permitting parents of these children to allow donation in this limited set of circumstances could meet much of the need for infant donors, and allow some good to come from a tragic loss.

In addition, the process the infant would undergo differs little from withdrawal of life-support, from retrieval of organs from patients declared dead by neurological criteria, or from non-heart-beating cadaver donation protocols, which are now ethically and legally acceptable.[4] In all these circumstances, the legislature has recognized the importance of preserving the viability of organs that will be donated, and has made provisions to allow the best chance for successful donation. As with these similar donation circumstances, medicine would find a humane way to accomplish the donation once an anencephalic infant was declared legally dead.

The situation of anencephaly is sufficiently unique that slippery slope problems can easily be avoided by this naturally self-limiting special case. The comparisons that opposing counsel draws to senile patients and developmentally delayed patients are as flawed as they are inflammatory. Physicians should be allowed in these special and limited circumstances to honor Mr. and Mrs. Doe's and other parents' choices to bring life and hope from otherwise tragic circumstances. For the reasons stated in brief and in oral argument, Mr. and Mrs. Doe respectfully request that the decision of the lower courts be reversed.

Oral Argument on Behalf of Hospital

May it please the Court. My name is Susan Smith and I represent Wake County Hospital in this matter.

There is no question that Baby Jane Doe's parents have shown great humanity, compassion, and concern for others in the face of their tragedy and sadness. But that is not the issue before this Court. The issue for this Court is whether we can harvest the organs of a child who is biologically alive. In seeking to have their child declared legally dead, the parents are asking this Court to go beyond the generally accepted definition of death, beyond the policy of the Dead Donor Rule of the Uniform Anatomical Gift Act,[5] and beyond the Wake statute, to create an additional common law standard regarding death and transplantation.

The hospital accepts the facts as stated by Baby Jane Doe's counsel, but would like to highlight additional medical facts:

First, an anencephalic birth is a live birth. Under the Dead Donor Rule, an individual is dead if she has sustained either irreversible cessation of circulatory and respiratory functions or irreversible cessation of all functions of the entire brain, including the brain stem. When she was born, Baby Jane Doe was alive in the sense that she was separated from the womb and capable of breathing and maintaining a heartbeat independent of her mother's body. Therefore, she did not meet the criteria of the Dead Donor Rule.

Second, anencephalic infants may show spontaneous movements of the extremities, startle reflexes, and pupils that respond to light. Some may show feeding reflexes, may cough, hiccup, or exhibit eye movements, and may produce facial expressions during the sleep-wake cycle. Moreover, anencephalic infants may be able to feel pain. They may reflexively avoid painful stimuli where the brain stem is functioning and thus are able to command an innate unconscious withdrawal reflex. Some physicians would maintain that this indicates pain recognition: a reaction perceived as signaling consciousness.

Third, anencephalic infants are rarely suitable for transplants. Anencephaly is often, though not always, accompanied by defects in various other body organs and systems, some of which may render the child unsuitable for organ transplantation. Most of the time, this genetic anomaly begins and can be diagnosed in the first trimester. Therefore, given advances in prenatal screening, the prevalence of this condition is decreasing. Creating an exception to harvest these babies' organs is not going to end, or even put a dent in, the organ shortage crisis.

Turning to my first argument, Baby Jane Doe is alive and her organs should not be harvested until she is allowed to die a natural death. What the parents are asking here is that we describe a group of people as suitable for termination in order to help another group of people deemed suitable for the continuation of life. In other words, killing may sometimes be a justifiable necessity for procuring transplantable organs. Such a utilitarian argument must logically fail. Let's look at two scenarios: what if Baby Jane Doe, instead of anencephaly, had a life expectancy of 50 years – would we kill her to harvest her organs for someone with a longer life expectancy of 80 years? What if she had a life expectancy of 20 years? What about five years? What about one year? What about one month? What about one day? Obviously it is not morally correct to say that because Baby Jane Doe was in the process of dying, like the rest of us, she did not have the right to live out her natural life.

Turning to my second argument, applying the Does' utilitarian argument opens up the dangerous possibility of saying that killing for the purpose of harvesting organs is justifiable. Where would we stop? Such a decision would open up the door to kill other humans who have similar cognitive impairments. In situations where hospitals have experimented with early organ harvest, there is evidence that doctors have wondered whether the organs of other severely cognitively impaired patients would be suitable for harvest and transplantation (Lafreniere & McGrath 1998). Would severely senile patients, patients in a persistent vegetative state, and severely retarded patients be placed in danger because they lack higher brain function?

In the past, only California has even considered a proposal to expand the Dead Donor Rule. That proposal was rejected. Moreover, Germany is the only country where anencephalic infants are considered legally dead. This decision has generally been criticized, and rejected elsewhere, however, because 1) it is factually incorrect to consider a person who is breathing and biologically active to be dead; 2) permanent loss of consciousness as a standard for legal death increases the fear that it might be applied to patients in a persistent vegetative state and raises the chance of misdiagnosis, resulting in wrongful death; and 3) the Dead Donor Rule is workable as written, because it avoids convoluted arguments about the definition of death and upholds the fundamental ethical principles that apply to the donation of organs.

Turning to my third argument, although Mr. and Mrs. Doe argue that this court should create a limited exception to the Dead Donor Rule for anencephalic infants, this court is not the appropriate forum for that decision. Changing the law and making policy are properly left to the Wake legislature. There is no basis here to expand the common law to equate anencephaly with death, according to either Wake law or any other commonly accepted definitions of death and suitability for organ donation, because:

- Even though the Wake statute is permissive in nature, its framers clearly did not intend to apply it to a situation where the individual is not being maintained on life support. The second clause, "where respiratory and circulatory functions are maintained by artificial means of support so as to preclude a determination that these functions have ceased," supports this interpretation;
- Anencephalic infants are not dead under the Uniform Determination of Death Act (National Conference of Commissioners on Uniform State Laws 1981), which requires an irreversible cessation of circulatory, respiratory or whole brain function; and
- Anencephalic infants do not qualify as dead under the Harvard Criteria (Report of the Ad Hoc Committee of Harvard Medical School 1963), which require unreceptivity and unresponsivity. Anencephalic infants have some responsiveness and may even withdraw from pain. In addition, the Harvard Criteria require no movement or breathing or reflexes. Anencephalic infants move and breathe on their own and have reflexes.

If we want to create a new exception to the Dead Donor Rule, the appropriate forum is the legislature. Legislative modification—the adoption of a statute to supplement or supplant the common law on legal death in these circumstances—could include public hearings through which members of the general public would both become more familiar with the issues and have their views taken into account in the framing of policy.

Legislators are free to explore the full range of public views and expert opinion when making their decisions. The views of groups representing patients, physicians, religious bodies, and the general public would be heard, because unlike the more limited and formal judicial process, the legislative process is designed to accommodate the full range of views.

Do we really want people to be dead solely for the purpose of transplantation when we would not consider them to be dead otherwise? If this court determines that anencephalic children should be organ farms, it will further erode public confidence in transplantation. On a matter so fundamental to society's sense of itself and so final for the individual involved, there should be a much greater medical and legal consensus than now exists before taking the major step to radically revise the concept of death.

For the reasons stated in brief and oral argument, the Wake County Hospital respectfully requests that this Court affirm the decisions of the lower court.

The Moot Court Process

Moot court, a simulated litigation experience that requires students to analyze, brief, and argue a case under a professor's supervision, is the one of the tools used by legal educators to teach argumentation to law students. During the moot court process, law students are presented with a fictional set of facts surrounding a legal case, often addressing a current issue of social or political importance. The hypothetical case and supporting documentation (the "Record") generally takes the form of a lawsuit that is being appealed from a trial court to an appellate court. The Record is balanced with both positive and negative facts for each side. Students are assigned to represent one of the parties, and must research and analyze a specific legal question (the "Issue") for their client. Their arguments on the Issue are presented to a mock panel of judges, first in a comprehensive written document, the appellate brief (the "Brief"), and then in an oral argument.

As advocates, students either ask the judges to adopt one legal rule over another, or argue for a favorable application of an existing legal rule. Because students are acting as attorneys during the moot court process, both the Model Rules of Professional Conduct and the potential threat of malpractice (Draisen 1996) require that they fully consider all legal arguments on behalf of their client, regardless of their personal views.

Written briefs that set forth the various legal arguments and authorities that support the client's position are submitted to the mock panel in advance of oral arguments. Then, during the oral arguments, student advocates must answer the judges' questions about the case and its potential implications. In the last phase of the moot court process, students switch sides and are required to deliver an oral argument on behalf of their previous opponent. Through the process of analyzing complex legal issues and constructing complete and persuasive arguments, law students enhance their ability to recognize and balance competing ideas and principles.

HOW AND WHY MOOT COURT ENHANCES ARGUMENTATION SKILLS FOR BIOETHICS DECISION-MAKING

Developing effective argumentation skills is essential to bioethics decision-making because it enables students to identify competing ethical and moral visions, to objectively evaluate various viewpoints and ultimately gain a nuanced, sophisticated understanding of the medical, legal, and social factors involved in the decision.

Argumentation is a process of logical reasoning to draw conclusions from the methodical presentation of arguments, discussion and debate (Soanes 1987). It is a primary method of addressing conflicting viewpoints about which ethical principles should take precedence and what course of action should be taken in a bioethics issue (Jennings 1999). Developing effective argumentation skills allows students to exchange ideas and have their ideas compete freely for recognition and adoption (Walton 2002).

Moot court hones argumentation skills because it demands preparation, organization, flexibility, and the ability to think quickly and respond convincingly when questioned (Moscovitz 1995). Thus, it works well to teach bioethics because legal analysis is consistent with principle-based analysis of bioethical issues (Furrow et al. 2008). In traditional bioethical analysis, the consistent set of competing bioethics principles – autonomy, justice, beneficence, and nonmaleficience – is weighed by decision-makers in each situation (Richardson 2000). Likewise, in legal analysis, decision-makers are asked to weigh competing potential rules or competing applications of the same rule (Furrow et al. 2008). As Kopelman (2009, p. 263) notes, both bioethics and law are interdisciplinary, "seek[ing] to include many perspectives and areas of expertise."

Teaching students how to recognize and weigh competing arguments enables them to more fully explore these interdisciplinary socially and politically complex issues and to determine which principle(s) should take priority. As illustrated by the Terri Schiavo case,[6] bioethics issues typically cannot be resolved with bright-line rules because the ethical decision will differ depending upon individual circumstances in background, culture, economics, education, religion, or other social factors (Beauchamp & Walters 1999). Thus, one challenge in teaching bioethics is to teach students to overcome social and psychological barriers that prevent them from keeping an open mind, and to allow them to identify competing ethical and moral visions and to dispassionately evaluate opposing positions.

Specifically, socio-religious factors inherent in the nature of most bioethics issues can create barriers to a balanced assessment of opposing arguments. Strongly held personal or religious beliefs can make it difficult for students to consider and weigh contradictory viewpoints in a bioethics argument (Kopelman 2009). Although individuals generally look to their background, faith, and culture for guidance when confronted by bioethics topics such as life, death, reproduction, and medical treatment, excessive personal involvement may push students toward an assessment based primarily on their individual beliefs. For instance, in the sample problem, if religious beliefs lead students to conclude that the anencephalic infant's spirit would continue to live in the world through organ donation, and that without it, her divine purpose would be thwarted, they might have difficulty recognizing or accepting the validity of the hospital's ethical concerns about harvesting organs before the baby is considered dead under the law. Conversely, students of faiths such as Buddhism may take issue with organ donation because they believe that removing the organs may affect the process of rebirth (Furrow et al. 2008).

Further, students may find it difficult to objectively evaluate various viewpoints because of psychological barriers. We, as humans, tend to feel discomfort when holding two contradictory but logically consistent ideas simultaneously; thus, psychological theory proposes that some people will reduce their psychological discomfort by rationalizing their beliefs (Festinger 1957). Others may simplify decisions by creating absolutes instead of acknowledging ambiguity (McElroy & Coughlin 2010), or accepting and remembering only the facts and arguments that conform to one's worldview while rejecting and minimizing contradictory information (Baron 2000). In the Baby Jane Doe sample problem, for example, students opposed to the parents' desire to have their anencephalic baby declared legally dead may discount or be unduly dismissive of the possible significance of the parents' right to make an autonomous decision on behalf of their offspring, or the fact that by allowing these organs to be harvested many other infants might live.

In addition, both psychological and social barriers limit students' ability to identify points of agreement or consensus with the opposing side. This failure to see points of agreement could distort students' views of the most difficult or contested arguments. Such contradictory positions are often viewed as diametric opposites, rather than as positions that might share common goals or ethical foundations. In the abortion debate, for instance, if students were representing Planned Parenthood and Operation Rescue, their positions could seem fundamentally inconsistent. Both sides, however, would like to decrease the number of abortions. In the Baby Jane Doe sample, students who identify with the parents may be unable to see that the hospital is actually interested in allowing organ donation, but only where it is consistent with its understanding of ethical responsibilities, so as to protect itself from legal liability and maintain the public trust in organ donation.

Moot court addresses these psychological and social barriers by requiring students to advocate on behalf of a fictitious third party client. Because students tend to identify with their role as attorneys (Juergens & McCaffrey 2008), this motivates them to consider all potential arguments for the client, even if they do not personally agree. Requiring students to argue on behalf of a fictitious client protects them from having to defend or change their own personal beliefs. As a result, students are less likely to create internal psychological or social barriers to opposing views.

Moot court's use of a specific, fictitious client also provides an opportunity for students to use argumentation within the social context of an issue. Because of its social nature, a bioethics question cannot be fully analyzed without recognizing the human story that underlies each issue. Advocates who learn to engage the panel with a human story (Rideout 2008) and present sound, principle-based arguments are likely to be most persuasive (Chestek 2008). Especially when students are struggling to overcome barriers created by personal beliefs, zealous advocacy for clients, first on one side, and then on the other, which includes telling the clients' stories and placing those stories in context, can help them to recognize the legitimate visions and principles on each side of the argument (Johnson et al. 2009).

To illustrate, in the sample problem a concerned doctor has been asked by loving parents to declare their anencephalic daughter legally dead (despite her heartbeat and respiration), so they can donate her organs to other babies who have a chance of survival. The choice the doctor faces demonstrates the real-life difficulty of harmonizing (DeMarco & Ford, 2006) abstract, contradictory interests between the rights of the individual and the broad, shared public health interest in increasing the number of organs available for transplantation. For a student who is personally resistant to either the doctor's position or the parents' position, considering the opposing client's story will make it harder to dismiss the opponent's arguments as wholly unsupported or completely unethical. Thus, this humanizing process is an essential part of argumentation, especially in the bioethics context, because bioethical decisions affect essential elements of human life and can have far-reaching effects on society.

LIMITATIONS AND SOLUTIONS TO THE USE OF MOOT COURT TO TEACH BIOETHICS DECISION-MAKING

Like any curricular tool, however, the moot court process may have some limitations when used in a different educational context, including the competitive nature of the moot court process, the procedural setting of the lawsuit, reliance on precedent in making legal decisions, and the two-party framework of legal issues. These potential barriers are easily surmountable with advance planning.

First, students participating in the moot court process, even when it is a component of an academic course, may view it solely as a competition to identify the best advocate, which, in turn, can create the perception that winning or coming up with "clever" arguments (Kozinski 1997) is more important than developing and presenting well-supported, substantive arguments. If the students' focus is on winning personal praise, then they may be less focused on persuading the moot court panel to adopt a position enabling their side to win. If misplaced emphasis on style, technique, and charisma appears to help students win, they are likely to deemphasize the importance of substantive arguments. Merely clever advocates, however, are unlikely to persuade a panel if they cannot discern which facts or authorities are important, or fail to supply thoughtful responses to oral argument questions (Martineau 1985).

One approach to minimize the perceived personal competition and maximize the importance of fully developed arguments is to recognize the best substantive argument, not the best presentation (should the winning substantive argument not be delivered by the student with the best lawyerly techniques) (Kozinski 1997). Alternatively, the scoring or grading system could be modified to more heavily weight the substantive points delivered by the student. In this way, the moot court experience could better reinforce not only the advocacy skills that are relevant in all legal arguments, but also the multifaceted analysis necessary to fully explore the social context and competing principles which are essential in bioethical analysis.

A second possible limitation of moot court in a bioethics context is the procedural setting of typical problems. The applicable procedural rules of the court, as well as the specific facts of the hypothetical case, may serve to limit the scope of the arguments students raise. These rules create the possibility that some aspects of the broader bioethical issue will not be raised because they were not triggered by the particular facts in the Record or certified by the court.

To overcome this procedural limitation, the professor should ensure that all students argue before a "hot bench" where the panel asks each advocate numerous challenging questions, particularly questions about the ramifications of the decision for society and future lawsuits. In contrast to a "cold bench" that allows the advocate to make a largely uninterrupted presentation to the court, a hot bench responds immediately to the weaknesses or untested assumptions underlying the advocate's position. Often, a hot bench includes judges that appear to lean toward opposite sides in the argument. A hot bench thereby requires students to elaborate on their positions, acknowledge and fully address the other side's arguments, and discuss the broader social implications of a decision for either side.

For instance, in the Baby Jane Doe sample, a hot bench might ask students representing Mr. and Mrs. Doe to address several possibilities: that 1) allowing a determination of legal death for their daughter would lead to similar but unwanted declarations for other patients with extremely short life expectancies; 2) hospitals might deny care to other anencephalic babies because they would be viewed as already dead; or 3) the public would fear that other cognitively impaired individuals would not receive care if physicians believed that they would be good candidates for organ donation. The Does' advocates might further be asked larger policy questions, such as whether they advocate abolishing the dead donor rule in favor of basing organ donation solely on the ethical ideals of respect for persons and nonmaleficence.

On the other hand, a hot bench might point out to the advocate for the hospital that all parties agree that Baby Jane Doe had no chance of survival and that Mrs. and Mrs. Doe wanted to donate their daughter's organs to salvage some good from their tragic loss. In light of these facts, a hot bench might ask the hospital's advocate to: 1) justify the loss of life-saving donor organs for children who might die without them, especially given the small pool of infant donors; 2) explain the refusal to draw a distinction between an infant with no possible chance of survival because of an absence of most of the brain and a person with a complete brain who has some small chance of recovery and survival; 3) explain the differences between declaring death in the case of anencephalic infant and the orchestrated deaths in non-heart-beating cadaver donation protocols, or 4) advise the court on resolving the potential conflict between the hospital's desire to promote consistency and avoid liability and the transplant physicians' desire to provide life-saving organs for children who need them. Advocates who expect this type of question will prepare more thoroughly and think more deeply about the parties' interests and the real-world consequences of their proposed outcomes.

The third possible limitation on effectively employing the moot court process in the bioethics context is the legal system's reliance on precedent (Kozinski 1993). Unlike bioethics arguments, which consider myriad sources, traditional legal argument relies almost exclusively on prior court decisions and enacted law. Attorneys rarely use legal commentary, scholarly articles, or non-legal sources of authority in their arguments (Coughlin, Malmud & Patrick 2008). As a result, moot court participants may initially be inclined to dismiss as irrelevant bioethics authorities that fall outside the usual scope for court decisions. Appellate courts, however, do have somewhat more discretion to consider policy arguments as well as arguments based upon non-legal sources, because appellate courts need not follow decisions made by trial courts and because policy affects how judges read precedent (Kozinski 1997).

To allow the greatest range of authority and arguments, the hypothetical lawsuit could be set before the Supreme Court of the U.S., where the Court is not constrained to follow its own decisions or cases decided by lower appellate courts (Kozinski 1997). This setting would allow non-legal considerations to be explored more fully, especially if the legal issue is one of first impression (where no prior case law exists on the Issue) or where the lower courts are divided on how to resolve the Issue.

Another possible solution is to modify the setting of the case from a court to another relevant decision-making body, such as a hospital ethics committee, a governmental or scientific advisory committee, or even an Institutional Review Board (which oversees experimentation with human subjects), where students could draw on more current and varied sources of authority. Medical journals, ethics scholarship, and opinions of other non-judicial bodies would provide additional relevant context for the issue. The ability to recognize the value different bodies are likely to place on various authorities and to craft arguments based on the most persuasive sources will not only hone analytic skills, but also illustrate the importance of considering audience.

A fourth possible limitation of the moot court process is the two-party structure. Because moot court problems are designed to simulate appellate court cases, each problem presents only two sides. In contrast, bioethics problems are multifaceted, and typically involve many individuals whose interests in the outcome of any particular decision may not be aligned. For instance, in the Baby Jane Doe case, the two sides presented are the parents and the health care providers, specifically the hospital. However, the interests of Baby Jane Doe may or may not be different from her parents', as there is a legitimate argument that the parents' desire to donate her organs may not be in Baby Jane Doe's best interest. Additionally, the various physicians involved, while sharing the hospital's concern about liability, may have different interests from the hospital or even from each other. For example, the obstetrician may or may not have the same interests at stake as the transplant surgeon. Further, the families of the children who would potentially receive the organs would also have a stake in the outcome of the Baby Jane Doe decision. In an adversarial, two-party structure like that of moot court, students might lose sight of the different interests that could be affected by the case and the variety of stakeholders who may want to have a voice in the decision-making process.

One possible solution could be to assign students to write or read amicus briefs in support of the additional interests at stake. Amicus briefs are submitted to the court from interested groups or individuals who are not directly involved in the litigation. These briefs generally argue for a particular outcome because it would promote a larger concern of the group, or would answer a question that is related to the particular controversy. Students who were required to read amicus briefs, or even to write them, would gain a greater appreciation for the additional parties that would be affected outside of the two that are directly involved in the litigation.

A hot bench, as discussed previously, would also help to highlight the various interests that are not formally represented in a moot court problem. A hot bench is more likely to ask the students questions about the potential consequences of the position the advocate has taken. If judges are not limited to questions about the specific case, but are instead encouraged to ask advocates questions about the wider implications on a variety of parties, students will have to think through all of the parties that would have an interest in the outcome, and determine how each would be affected by the decisions. These types of questions would encourage comprehensive thought about the entire problem and complete analysis of the wider impact of any ruling.

For the reasons described above, moot court can be an effective tool to promote effective argumentation and to teach students the importance of social context in the bioethics decision-making process. There are many ways to implement moot court to fit with curricular needs. Like all pedagogical changes, however, the ideas are only limited only by the professor's creativity.

NOTES

1. The authors would like to thank Wake Forest University School of Law students Stephen Bell, Alayna Ness, Beau McNeely, and Ron Payne for their assistance with researching and editing this chapter.
2. The authors would like to acknowledge and thank Professor Mark Hall, Fred D. and Elizabeth L. Turnage Professor of Law at Wake Forest University School of Law and co-founder of the Wake Forest University Center for Bioethics, Health and Society, who initially authored the case file for the 2006 Bioethics and Law class, as well as the oral argument on behalf of the Does. Additional documents related to this case file are available upon request. Please contact Professor Chris Coughlin, coughlcn@wfu.edu.
3. Arguments used in the oral argument were primarily obtained from Truog and Robinson (2003) (hereinafter "Truog"); President's Commission for the Study of Ethical Problems in Medicine and Biomedicine and Behavioral Research (1981); and Lafrieniere and McGrath (1998).
4. This is an "orchestrated death" for purposes of organ transplant where the patient's ventilator is withdrawn and the doctor watches for the heartbeat to stop. The organs are harvested for transplantation after the patient's heartbeat stops (usually within two to five minutes) and cardiac death rather than brain death occurs. See Koostra et al. (1995).
5. The Dead Donor Rule from the Uniform Determination of Death Act (1981, section 1) states: "An individual who has sustained either (1) irreversible cessation of circulatory and respiratory functions, or (2) irreversible cessation of all functions of the entire brain, including the brain stem, is dead. A determination of death must be made in accordance with accepted medical standards."
6. Quill (2005) and Hyde and McSpirit (2007) both discuss different aspects of this case. Terri Schiavo was a Florida woman who, after cardiac arrest in 1990, was in a persistent vegetative state without cortical function for fifteen years until her death on March 31, 2005. Her parents opposed a petition by Ms. Schiavo's husband to withdraw her feeding tube and allow her to die. On March 21, 2005, the U.S. Congress enacted a bill, signed by then-President Bush three days later, transferring jurisdiction of the case from the Florida state courts to a federal district court for review of the Florida courts' decisions granting the husband's request. The federal courts, including the U.S. Supreme Court, rejected or refused to hear the parents' petition.

REFERENCES

Baron, J. (2000). *Thinking and deciding* (3rd ed.). New York: Cambridge University Press.

Beauchamp, T. & Walters, L. (1999). *Contemporary issues in bioethics* (5th ed.). Belmont, CA: Wadworth.

Chestek, K. (2008). The plot thickens: The appellate brief as story. *Journal of the Legal Writing Institute* 14: 131.

Coughlin, C., Malmud, J. & Patrick, S. (2008). *A lawyer writes: A practical guide to legal analysis.* Durham, NC: Carolina Academic Press.

Demarco, J. & Ford, P. (2006). Balancing in ethical deliberation: Superior to specification and causistry. *Journal of Medicine and Philosophy* 31: 493.

Draisen, D. (1996). The model rules of professional conduct and their relationship to legal malpractice actions: A practical approach to the use of the rules. *Journal of the Legal Profession* 21: 81.

Festinger, L. (1957). *A theory of cognitive dissonance.* Evanton, IL: Row Peterson.

Furrow, B., Greaney, T., Johnson, S., Jost, T. & Shwartz, R., (2008). *Bioethics: Healthcare law and ethics* (6th ed.). Eagan, MN: Thomson West.

Hyde, M. J. & McSpirit, S. (2007). Coming to terms with perfection: The case of Terri Schiavo. *Quarterly-Journal of Speech* 93: 150–178.

Jennings, B. (1999). The liberal neutrality of living and dying: Bioethics, constitutional law, and political theory in the American right-to-die debate. *Journal of Contemporary Health Law and Policy* 16: 99.

Johnson, S., Krause, J., Saver, R. & Wilson, R. (2009). *Health law and bioethics: Cases in context.* New York: Aspen Publishers.

Juergens, A. & McCaffrey A. (2008). Roleplays as rehearsals for "doing the right thing" – adding practice in professional values to moldovan and united states legal education. *Washington University Journal of Law and Policy* 28: 164.

Koostra, G., Daemen, J. H. & Oomen, A. P. (1995). Categories of non-heart-beating donors. *Transplantation Proceedings* 27(5): 2893.

Kopelman, L. (2009). Bioethics as public discourse and second-order discipline. *Journal of Medicine and Philosophy* 34: 261–273.

Kozinski, A. (1993). What I ate for breakfast and other mysteries of judicial decision making. *Loyola of Los Angeles Law Review.* 26: 993.

Kozinski, A. (1997). In praise of moot court – not! *Columbia Law Review* 97(1): 178–197.

Lafreniere, R. & McGrath, M. (1998). End of life issues: Anencephalic infants as organ donors, *Journal of the American College of Surgeons* 187: 443–447.

Martineau, R. (1985). Moot court: too much moot, and not enough court. *Fundamentals of Modern Appellate Advocacy.* San Francisco: Bancroft Whitney. Reprinted from American Bar Association Journal 67, October 1981: 1294–1297.

McElroy, L. & Coughlin, C. (2010). The other side of the story: Using graphic organizers as cognitive learning tools to teach students to construct effective counter-analysis. *University of Baltimore Law Review* 39: 227–253.

Moskovitz, M. (1995). *Winning an Appeal* (3rd ed.). Charlottesville, VA: Michie Butterworth.

National Conference of Commissioners on Uniform State Laws (1981). Uniform Determination of Death Act.

President's Commission for the Study of Ethical Problems in Medicine and Biomedical and Behavioral Research (1981). *Defining death: Medical legal and ethical issues in the determination of death.* Available from: <http://bioethics.georgetown.edu/pcbe/reports/past-commissions> Accessed 13 June 2011.

Quill, T. (2005). Terry Schiavo – a tragedy compounded. *The New England Journal of Medicine* 352: 1630–1633.

Report of the Ad Hoc Committee at Harvard Medical School to Examine the Definition of Brain Death (1968). A definition of irreversible coma. *Journal of the American Medical Association* 205: 337–340.

Richardson, H. (2000). Specifying and interpreting bioethical principles. *Journal of Medicine and Philosophy* 25; 29.

Rideout, J. (2008). Storytelling, narrative rationality and legal persuasion in legal writing. *The Journal of the Legal Writing Institute* 14: 57.

Soanes, C. (Ed.) (1987). *The compact edition of the Oxford English Dictionary* (26th ed.). Oxford: Oxford University Press.

Truog, R. & Robinson W. (2003). The role of brain death and the dead donor rule in the ethics of organ transplantation. *Critical Care Medicin,* 31: 2391.

Walton, D. (2002). *Legal argumentation and evidence.* University Park, PA: Penn State University Press.

Weinstein v. Bradford, 423 U.S. 147 (1975).

Part II

Moral Relationships and Responsibilities

4 Dignity Can Be a Useful Concept in Bioethics

Rebecca Dresser

People outside the field may find it surprising, but human dignity is a contested concept in bioethics. The dignity controversy is a relatively recent development that began just a few years ago. Until that time, people addressing bioethics issues sometimes invoked the concept of human dignity, but it was neither a central focus of nor a controversial topic among scholars in the field.

In 2003, however, an argument over dignity commenced. That year, the *BMJ* (formerly known as the *British Medical Journal*) published an article by the respected bioethics scholar, Ruth Macklin. In her article, "Dignity Is a Useless Concept," Macklin criticized the ways that scholars and policymakers have used this ethical concept in bioethics analysis (2003). According to Macklin, dignity "is a useless concept in medical ethics and can be eliminated without any loss of content" (p. 1420).

Macklin described several problems with scholarly and policy uses of dignity. One is that, in her view, "appeals to dignity are either vague restatements of other, more precise, notions or mere slogans that add nothing to an understanding of the topic" (p. 1419). In the worst cases, Macklin asserted, writers use dignity as a slogan to avoid the harder task of supplying substantive arguments for their positions. In other cases, she wrote, appeals to dignity are simply redundant, adding nothing to an analysis. For example, she argued that in the end-of-life context, appeals to dignity simply duplicate appeals to the principle that patients should have control over decisions about life-sustaining treatment. As another example, she cited the Council of Europe's convention for the protection of human rights and dignity of the human being with regard to the application of biology and medicine (Council of Europe 1997). In that document, appeals to dignity seemed to Macklin "to have no meaning beyond what is implied by the principle of medical ethics [called] respect for persons," such as the need to obtain informed, voluntary consent; the requirement to protect confidentiality; and the need to avoid "discriminatory and abusive practices" (p. 1419).

Macklin's article provoked many responses, which in turn provoked still more commentary, with some writers defending and others challenging Macklin's assertions. For my part, I appreciate Macklin's challenge to think more carefully about what dignity can mean, should mean, and can contribute to bioethics analysis. But I disagree with her position, for several reasons.

First, the concept of dignity plays a role in many areas of moral and political discourse—why not in bioethics? And in analyzing bioethical issues in human subject research, end-of-life care, apologies for medical errors, and other topics, writers often see dignity as a relevant and useful analytic concept. Judging from the bioethics literature before and after Macklin's challenge, many writers see something valuable in the concept of dignity. If people inside and outside the field believe the concept has meaning and significance, shouldn't we be open to the possibility that they are correct?

Second, I think Macklin applies a double standard in her demand to reject dignity on grounds of imprecision or inadequate explication. Other basic concepts in bioethics are subject to differing interpretations and are sometimes used as slogans—respect for persons and autonomy often suffer this fate. Quality of life and futility are other examples of contested concepts in bioethics. Of course, we should as scholars define, defend, and justify the concepts we see as central to bioethics analysis. But why must we eliminate dignity and not other contested concepts? It seems to me that discussing, indeed, arguing about, bioethical concepts is essential to advancing our understanding of what these concepts can mean and when they should and should not influence ethical and policy judgments. As Timothy Caulfield and Audrey Chapman put it, "by using the concept of dignity as an avenue for exploring different philosophical approaches, we promote transparency, encourage dialogue, and help to avoid the simplistic application of dignity . . ." (2005, p. 738). From this vantage point, what we need is more conversation, not silence, about dignity in bioethics.

Third, even if Macklin is correct in her judgment that dignity analyses have so far been inadequate, why is it time to abandon the effort? Is there a statute of limitations on permissible inquiry into ethical concepts? Why should those interested in the concept's relevance to bioethics be discouraged from examining it? The concept of dignity and its application to bioethics issues has not yet been subjected to the same degree of analysis and scrutiny as have concepts like autonomy and beneficence. With further attention from scholars, clinicians, and policymakers, dignity might turn out to offer more than what Macklin thinks it does.

Fourth, Macklin incorporates a reductionist approach to bioethics that may be useful in some contexts, but can be arid and unsatisfying in others. I don't believe that we should always aim to distill into basic principles our ethical analyses of complex problems in medicine and science. Why not engage in more fine-grained inquiry, with the possible insights such inquiry can deliver? I think we can learn a lot from writers adopting "thick" concepts like dignity. Such concepts may overlap with other moral concepts, but at the same time can reveal details and dimensions that might otherwise go unnoticed (Ashcroft 2005).

In the remainder of this chapter, I will defend this position by presenting examples of dignity talk in public and popular discourse, as well as in a few bioethics contexts. Then I will consider dignity from the perspective of a patient with serious illness. From both professional and personal perspectives, I see dignity as a concept that can illuminate and enrich bioethical analysis.

DIGNITY IN CONTEMPORARY DISCOURSE

After I was asked to speak about dignity at the "Bioethics, Public Moral Argument, and Social Responsibility" conference, I began looking for dignity language and located it in many places. I found that dignity is commonly invoked in contemporary moral discourse. Below, I describe selected examples of dignity talk reported in the *New York Times* during the winter of 2008 and spring of 2009. Then I present a few examples of helpful and not-so-helpful uses of dignity in bioethics discourse.

President Barak Obama referred to dignity twice in his inauguration speech.

In addressing our nation's place in the world, he said, "Know that America is a friend of every nation and every man, woman and child who seeks a future of peace and dignity, and that we are ready to lead once more" (Transcript 2009, p. 2). And in discussing his plans for government initiatives in health care and other social programs, he said, "The question we ask today is not whether our government is too big or too small, but whether it works—whether it helps families find jobs at a decent wage, care they can afford, a retirement that is dignified" (p. 5).

Another example comes from U.S. Supreme Court Justice Ruth Bader Ginsburg. In a 2009 speech, she defended her view that U.S. judges should consider the rulings of foreign courts when those rulings offer wisdom and good reasoning (Liptak 2009). As an example of an "eloquent" and "persuasive" foreign ruling, she cited a decision by the Supreme Court of Israel that prohibited the use of torture to elicit information from suspected terrorists. She took from that opinion the message "that we could hand our enemies no greater victory than to come to look like that enemy in our disregard for human dignity" (as cited in Liptak 2009, p. 2).

As a law professor, I can tell you that dignity appears often in legal documents and commentary. Besides its significance in human rights analysis (Hayry 2004), dignity is often invoked as a value underlying basic legal protections and civil liberties. For example, a news article in December 2008 described the work of the American Constitution Society, a group of liberal and progressive lawyers and law professors (Savage 2008). The group's mission is to ensure that "human dignity, individual rights and liberties, genuine equality and access to justice enjoy their rightful, central place in American law" (as cited in Savage 2008, p. 2).

Dignity talk is present in the arts, too. For example, it appeared in a March 2009 film review entitled, "The High Cost of Dignity: Recalling the Troubles in Stark Detail" (Scott 2009). *Hunger* is a drama about the prison in which British officials confined members of the Irish Republican Army during the conflict in Northern Ireland. The film shows how prisoners used hunger strikes and other forms of protest to assert the political nature of their offenses. In his review, the critic described how the film sought "to bring human dignity into a place where it ha[d] all but vanished" (p. 1).

Dignity appeared as well in a 2008 news feature on the negative consequences of "elderspeak, the sweetly belittling form of address that has always rankled older people" (Leland 2008). A reporter told the story of a 61-year-old woman who had been given a special packaging system for her cancer medication. Her pharmacists assumed that she would not take the pills properly, due to her age. But before this, she had been fine on her own, taking the pills as directed. Moreover, the device was poorly designed and she was unable to open it. She was angry and insulted, she said, because, "I believed my dignity and integrity were being assaulted" (as cited in Leland 2008, p. 2).

My aim in relaying these reports is to show that dignity is a concept that has meaning and significance for both ordinary and extraordinary people addressing contemporary moral and social issues. This includes people considering ethical issues in medicine and research. Below I present examples of scholarship in which I think writers have invoked dignity in ways that enhance their bioethical analyses.

One such example comes from physician Aaron Lazare's work on apology in medical practice. In a compelling article, Lazare described how apologies can play a healing role when clinicians commit offenses against patients (2006). Lazare offered a long list of such offenses, including "excessive waiting time, failure to address the patient by his or her preferred name, violations of privacy of conversations and records, inappropriate body exposure of the patient, failure to listen to the patient and adequately explain the nature of illness or procedures, inadequate communications among the treatment team, and making disparaging or condescending comments about the patient's medical conditions or habits" (p. 1403). Apologies for such offenses can restore the patient's self-respect and dignity, Lazare wrote, for "the patient often experiences the offense as a humiliation that he or she may express with words, such as 'I was treated with disrespect' or 'as dirt' or 'less than a person' or 'like I did not matter'" (p. 1402). In such cases, Lazare contended, "the healing factor must be the restoration of dignity, which may take the form of the physician humbling himself or herself" (p. 1402).

Here I think Lazare used dignity effectively to capture a rich sense of the insults and injuries that patients endure, and the clinician's ethical responsibility in such cases. It would be possible to talk about these problems as violations of autonomy or beneficence, but such an approach would lack the emotion and power that make Lazare's analysis so effective.

Recognizing dignity as a moral concern can also explain why a general principle like respect for persons is important. An example comes from work by Paul Appelbaum and Charles Lidz on the therapeutic misconception in research (2006). The therapeutic misconception arises when patients enroll in research without understanding important facts about clinical trial participation (Henderson et al. 2007). Methodological restrictions necessary to produce valid scientific data, such as randomization and blinding, make study participation different from the clinical care patients receive from their personal physicians. In explaining why the therapeutic misconception is an ethical concern, Appelbaum and Lidz turn to dignity language.

According to Appelbaum and Lidz, when patients enroll in trials without understanding how research participation differs from clinical care, researchers have harmed patients' "dignitary interests" (p. 369). The harm occurs, these scholars believe, whether or not there are any harmful effects on the patient's health outcome. It is generally agreed that patients should have the opportunity to decide for themselves whether or not to participate in the knowledge-seeking mission of research. According to Appelbaum and Lidz, when a researcher fails to make patients aware of the differences between ordinary medical care and clinical trial participation, the researcher regards patients as useful data-sources, rather than as persons owed the freedom to make important choices about their lives. Appelbaum and Lidz see this as a violation of the patients' dignity, whether or not it harms them in other ways. Here again, I think dignity language supplements and enriches our understanding of why certain practices in medicine and research raise moral issues.

Admittedly, dignity language is not always put to good use in bioethics debates. For example, the concept is used to support opposing positions on abortion and physician-assisted death. Those supporting the pregnant woman's control over her body characterize abortion restrictions as violations of her dignity (Siegel 2008), while opponents of abortion cite the inherent dignity of human life at any developmental stage as the basis of their position (Ashcroft 2005). Similarly, those favoring legalization of physician-assisted suicide and active euthanasia invoke "a right to die with dignity," while their opponents speak of such measures as threatening the dignity of the vulnerable and incapacitated (Schroeder 2008; Sheehan 2008).

As I said earlier, dignity talk is not always useful or enlightening and it is both fair and necessary to point out when that occurs. But sometimes dignity enhances our understanding, and we would lose something important if we banished it from bioethics.

A PERSONAL PERSPECTIVE ON DIGNITY

A place where I think dignity can be especially useful is in understanding the ethics of caring for patients with serious illness. Psychiatry professor Stanley Giannet conceives of dignity in medical care as "a psychospiritual connection that involves empathy, connection, and compassion" (2003, p. 11). And Caulfield and Chapman point to a general consensus that "human dignity is tied to notions of human worth" (2005, p. 737). In this section, I discuss how dignity and indignity can be experienced when one is a patient—how, in the words of Caulfield and Chapman, human worth can be expressed and perceived by patients in the medical setting.

A few years ago, I was diagnosed and treated for oral cancer. I learned a lot about medical ethics from that experience, and some of what I learned was about how important dignity can be to a patient. In a previous essay, I wrote about this topic, and there I described how dignity concerns may be connected to privacy, communication, personal knowledge, and dependence (Dresser 2008). Below I repeat and expand on what I said in that essay.

Dignity concerns arise in connection with the broader area of personal privacy, but contrary to Macklin, they go beyond protecting the confidentiality of medical communications. Patients enter a world of forced and one-sided intimacy with strangers. Patients feel a loss of dignity when care intrudes into areas raising particular privacy concerns. Having to wear "flimsy and revealing hospital gowns" and being gossiped about by staff are some of the violations that can occur (Capron 2003).

Being wheeled through the halls in a wheelchair or on a gurney feels undignified, too, especially when the halls are public places with visitors who cannot resist staring at the sight. Having medical students and residents troop into a hospital room to make one an object of study can also be experienced as an indignity.

The outward signs of illness create a heightened need to be treated in a dignified manner. Hair loss, severe weight loss, and other unwelcome changes make patients sensitive about appearing in public. Small actions, such as making eye contact with a skinny, bald cancer patient, are ways to confer dignity on such patients. It is all too tempting to look away from people who are obviously ill, in part because they are reminders of human frailty and mortality. Yet to maintain a patient's sense of self-worth, it is essential for those in "the kingdom of the well" to establish a connection with those in the "kingdom of the sick" (Sontag 2001, p. 3).

Another dimension of dignity is respect for the patient's personal knowledge. Being diagnosed with a life-threatening illness is life-altering. Priorities, relationships, and social roles undergo drastic change. Patients face mortality in ways that healthy people cannot imagine. In this sense, patients know more than the relatives, friends, and clinicians around them. Many patients suffer through chemotherapy, radiation, surgery, and other burdensome interventions, and this demands a kind of strength never before required. Dignity is promoted when others honor the patient's ordeal and recognize the person who is enduring the assaults of illness and treatment.

Serious illness also brings a new kind of dependence, and being dependent feels undignified to many people. But some clinicians respond to a patient's desire for independence by giving her too much responsibility for decisions she is unequipped to make. Telling a patient that it is up to her to decide whether her symptoms merit hospitalization is inappropriate when the patient lacks the medical expertise needed to make such a choice. On the other hand, pressuring patients to accept beneficial treatments they are resisting can be appropriate.

Clinicians and informal caregivers respect human dignity when they attend to patients' needs for help in navigating the complicated course of a serious illness. Thoughtless invocations of autonomy can conflict with patients' dignity interests.

Patients are persons deserving of high-quality care, and sometimes this requires others to assume or share with them the authority for making difficult medical decisions.

Two of my experiences as a cancer patient illustrate the real-life importance of dignity in medical care. One involves my diagnosis and the other a poor treatment decision I made. In each situation, nurses reached out to me in ways that acknowledged my dignity, my human worth.

"Breaking bad news" is a topic often covered in medical ethics courses (Poulson 1998) and one that I have often discussed with medical students. In 2003, I learned what it was like to be the recipient of such news. Before I received my cancer diagnosis, I had some symptoms that were dismissed by a couple of different clinicians. I was worried about the symptoms, but I wasn't that worried. So when I finally got to the right sort of expert and he said he was "pretty sure" it was a malignant tumor, I was shocked, so shocked that I almost fainted.

In what would become a pattern, the human connection came from a member of the nursing profession. While the doctors left me alone as they prepared to conduct a biopsy, a nursing student noticed my distress and asked if I would like some water. Meanwhile, the doctor and his resident seemed tense and uncomfortable, focusing on the biopsy rather than the patient before them. As I remember it, nothing was said at that point about treatment and prognosis. I spent a couple of days not knowing whether the cancer was treatable or terminal.

I'm happy to report that the cancer was treatable, but the tumor was advanced and the treatment regimen demanding. I had chemotherapy and radiation, which made eating difficult. The oncologists explained that eating could become a problem for me and even provided a PowerPoint slide show about the treatment regimen's likely side effects. They said that about 50% of patients undergoing this form of treatment required a gastrostomy feeding tube for adequate nutrition.

When I heard about this possibility, I resolved that I would not be in the group that needed a feeding tube. I knew about feeding tubes from teaching and writing about high-profile court cases involving patients like Nancy Cruzan and Terri Schiavo (Dresser 2007). I couldn't imagine needing the same sort of intervention that had been at the center of the controversies over their treatment. I also didn't want the physical intrusion associated with the tube. And I most definitely did not want to go into the hospital. I saw the hospital as a dangerous place where patients were exposed to infection and the possibility of a medical error.

As the days passed, however, I became dehydrated, lost a lot of weight, and was doing badly. It was one of the nurses at the chemotherapy center who finally persuaded me to change my mind. What convinced me was how worried she looked when she said, "you really need to do this." I finally realized that I was facing a serious health threat, and consented to the tube.

The feeding tube proved to be a blessing. My earlier refusal was a poor choice, even though I knew the relevant "facts" about the intervention. What I needed was not to be left alone to make my own choice. Instead, I needed someone to say, "You know, you have really lost a lot of weight. I know that you don't find this an appealing prospect, but I'm worried because it's affecting your health and perhaps ability to complete the treatment for your cancer." With her statement of concern, the nurse recognized my vulnerability and helped me to reconsider my earlier decision.

In these two situations, my freedom to make treatment choices was respected (perhaps too respected) and my confidentiality was not violated. I was not subjected to "discriminatory and abusive practices." The clinicians caring for me observed all of the requirements Macklin included in her description of the respect for persons principle (2003, p. 1419). But some of the clinicians failed to recognize, in Caulfield's and Chapman's words, my "human worth" (2005), my needs as a particular individual reacting to an overwhelming situation. The doctors making the diagnosis were competent, skilled technicians, but they were not able to respond to my distress in the meaningful way that the nursing student could. She saw me as someone whose life had just been turned upside down and made the human connection that helped me to cope. And in the feeding tube situation, it was again a nurse who saw "the genuine patient and her circumstances" (Mayer 2004), the nurse who treated me with dignity by caring enough to disagree with my autonomous choice.

CONCLUSION

Philosopher Doris Schroeder had it right when she said, "dignity is a slippery idea, but it is also a very powerful one and the demand to purge it from ethical discourse amounts to whistling in the wind" (2008, p. 237). Rather than railing against the use of dignity in bioethical analysis, I urge colleagues to join me in considering its relevance to bioethics issues. In my view, a careful and thoughtful examination of dignity could point the way to a more nuanced understanding of our ethical responsibilities to patients.

NOTE

1. Portions of this chapter are based on my earlier essay, "Human Dignity and the Seriously Ill Patient" (Dresser 2008).

REFERENCES

Appelbaum, P. & Lidz, C. W. (2006). Re-evaluating the therapeutic misconception: Response to Miller and Joffe. *Kennedy Institute of Ethics Journal* 16: 367–373.

Ashcroft, R. E. (2005). Making sense of dignity. *Journal of Medical Ethics* 31: 679–682.

Capron, A. (2003). Indignities, respect for persons, and the vagueness of human dignity. *BMJ Rapid Responses*, December 31. Accessed online at http://bmj.bmjjournals.com/cgi/eletters/327/7429/1419. July 30, 2009.

Caulfield, T. & Chapman, A. (2005). Human dignity as a criterion for science policy.

PLoS Medicine 2: 736–738.

Council of Europe (1997). Convention for the protection of human rights and dignity of the human being with regard to the application of biology and medicine. Strasbourg: Council of Europe.

Dresser, R. (2007). Schiavo and contemporary myths about dying. *University of Miami Law Review* 61: 821–846.

Dresser, R. (2008). Human dignity and the seriously ill patient. In *Human dignity and bioethics: Essays commissioned by the President's Council on Bioethics.* 505–512. Available from: <http://www.bioethics.gov/reports/human_dignity/index.html> Accessed July 30, 2009.

Giannet, S. (2003). Dignity is a moral imperative. *BMJ Rapid Responses*, December 25. Available from: <http://bmj.bmjjournals.com/cgi/eletters/327/7429/1419> Accessed July 30, 2009.

Hayry, M. (2004). Another look at dignity. *Cambridge Quarterly of Healthcare Ethics* 13: 7–14.

Henderson, G. et al. (2007). Clinical trials and medical care: Defining the therapeutic misconception. *PLoS Medicine* 4: 1735–1738.

Lazare, A. (2006). Apology in medical practice: An emerging clinical skill. *Journal of the American Medical Association* 296: 1401–1404.

Leland, J. (2008). In "sweetie" and "dear," a hurt for the elderly. *New York Times,* October 7.

Liptak, A. (2009). Ginsberg shares views on influence of foreign law on her court, and vice versa. *New York Times,* April 12.

Macklin, R. (2003). Dignity is a useless concept. *BMJ* 327: 1419–1420.

Mayer, L. E. (2004). Clarifying the concept of dignity. *BMJ Rapid Responses,* February 5. Accessed online at http://bmj.bmjjournals.com/cgi/eletters/327/7429/1419. July 30, 2009.

Poulson, J. (1998). Bitter pills to swallow. *New England Journal of Medicine* 338: 1844–1846.

Savage, C. (2008). Liberal legal group is following new administration's path to power. *New York Times*, December 11.

Schroeder, D. (2008). Dignity: Two riddles and four concepts. *Cambridge Quarterly of Healthcare Ethics* 17: 230–238.

Scott, A. O. (2009). The high cost of dignity: Recalling the troubles in stark detail. *New York Times*, March 20.

Sheehan, M. N. (2008). Book review of *Human dignity and bioethics: essays commissioned by the President's Council on Bioethics*. *Journal of the American Medical Association* 300: 2926–2927.

Siegel, R. B. (2008). Dignity and the politics of protection: Abortion restrictions under Casey/Carhart. *Yale Law Journal* 117: 1694–1800.

Sontag, S. (2001). Illness as metaphor. In S. Sontag (Ed.), *Illness as metaphor and AIDS and its metaphors*. New York: Farrar, Straus and Giroux (Picador).

Transcript. (2009). Barack Obama's inaugural address. Available from: <http://www.npr.org/templates/story/story.php?storyId=99590481> Accessed July 30, 2009.

5 Appeals to Human Nature in Biomedical Ethics

Managing Our Legacies, Loyalties, and Love of Champions

Eric T. Juengst

Appeals to human nature have a strong and ancient history in moral philosophy, and in contemporary bioethics that tradition remains vibrant. In order to interpret our relationships with one another and the world around us, it makes good sense to begin with an understanding of what we cherish about being human. Doing so can, in theory, help identify features of the human experience we should always strive to preserve, and set aspirational goals for our actions with respect to the rest. In the practical contexts of bioethics and science policy, however, developing or adopting a universal, normative account of human nature as a prerequisite to decision-making is never an option, since such an account is perennially "in process." The best that these analysts can hope for is to snatch from the ongoing project of philosophical anthropology insights and ideas about particular aspects of human nature that can be helpful in their tasks at hand. In doing so, of course, they import into their policy-making deep, competing, and even incommensurable philosophical assumptions, on which no public or philosophical consensus may exist. This makes their arguments easy to demolish, on both logical and political grounds. So easy, in fact, that the real messages of such appeals are often swept aside with the debris, and lost.

The purpose of this essay is to poke about in the ruins of three bioethical and science policy appeals to human nature, in order to see what can be salvaged of value to public deliberations about the moral limits of biomedical interventions in human beings. First, I examine "life-cycle traditionalism" as it appears in arguments against the radical extension of the human life span through the (still hypothetical) interventions of "anti-aging medicine." When Dan Callahan characterizes our species' experience to date of growth, maturation, senescence, and death as one of the constitutive parts of human nature that should not be lost through biomedical intervention, is there anything more to this than reactionary privileging of the past? From there, I turn to the concerns over "species-altering experiments" that have led some to call for an international convention on the "preservation of the human species." When George Annas and Lori Andrews suggest that human rights are jeopardized by interventions that could change our biological identification as members of the human species, are they simply confusing two senses of the word "human," or pointing to something deeper? Finally, I examine the concept of "natural talents" as it is used in discussions of performance enhancement in sport. When Tom Murray worries that performance enhancement might undermine sports' celebration of the "virtuous perfection of natural human talents," is he inadvertently subscribing to an outdated "geneticism," or picking up on a feature of human experience worth cherishing?

I choose these three examples because each emphasizes a different dimension of our experience of ourselves as important to our collective self-understanding. The life-cycle traditionalists stress our historicity: "by nature" we are creatures with a past and enough memory to bind us to our origins. The species preservationists stress our incarnation: "by nature" we are biological creatures of a particular family, with all the loyalties and limits of family ties. The natural talent perfectionists stress our diversity one from another: "by nature" we are creatures that can be sorted into ranked subgroups for different purposes, so that our social structures tend toward hierarchies of perceived excellence in different spheres of practice.

To look ahead, what I find is that each of these aspects of human nature is distinctly double-edged. Each can be a source of core human values—but just as easily a source of human evil. Appreciating the significance of these dimensions in abstraction cannot help much in deciding which legacies to honor, which loyalties to respect, and which interpersonal stratifications to celebrate—i.e., in the work of practical ethics and policy-making. In fact, since the vices to which our legacies, loyalties, and classification schemes can fall prey are so much easier to agree upon than their virtues, it may be that the most useful practical role for appeals to human nature is a cautionary one. Where a biomedical intervention touches one of these dimensions of human nature is a signal that the moral stakes are high. But those stakes are not always about what might be lost from the human experience. There are also moral dangers in what might be perpetuated, if we allow the preservation of these hallmarks of human nature to overshadow our other values. Or so I will argue below.

LEGACIES AND "LIFE-CYCLE TRADITIONALISM"

One of the most common arguments against biomedical attempts to control, forestall, or reverse the aging process is that aging is part of the life-cycle that defines human beings, and tampering with that cycle could literally be "dehumanizing." This view is what Dan Callahan calls "life-cycle traditionalism":

> It is based on the biological rhythm of the life cycle as a way of providing a biological boundary to medical aspiration. This view looks to find a decent harmony between the present biological reality of the life cycle as an important characteristic of all living organisms and the feasible, affordable goals of medicine" (Callahan 1995, p.23).

Of course, the first immediate point to make is that while the life-cycle is "an important characteristic of all living organisms" for biologists, the "life-cycle" they have in mind when they make that claim is very different from the trajectory from birth and growth to senescence and death to which Callahan refers. The cycle that is important to biologists, after all, is the sexual life-cycle between the fusion of gametes and the developmental stages through which an organism matures to the point of producing its own gametes with which to repeat the process. In the biology classroom, one finds that none of the "life-cycle" charts—whether for plants, animals or humans—even include senescence and death as significant parts of life, since, from an evolutionary point of view, they are irrelevant. Whatever strength life-cycle traditionalism may have, in other words, it cannot be its affinity with a scientific worldview, as Callahan suggests above.

Others are more careful to avoid wrapping themselves in the mantle of natural science in this way, to stress that it is the role of senescence as a core part of human self-understanding that is important to preserve. As the President's Council asks: "Might we be cheating ourselves by departing from the contour and constraint of natural life (our frailty and finitude) which serve as a lens for a larger vision that might give all of life coherence and sustaining significance?" (President's Council on Bioethics, 2003)

Appealing to the "natural life-cycle" as a normative guide has interesting analogs within professional medical ethics, where it plays an important role in the debates on forgoing life-sustaining treatment. There, one often hears that a "natural death" is the proper goal of care for the terminally ill and that, quite apart from considerations of personal autonomy, extraordinary attempts to "prolong life" artificially can sometimes be demeaning and dehumanizing for patients. This argument assumes that there is a normative natural order to life's events and their pacing in individual human lives. Critics of anti-aging medicine can argue that the fact that we can see the importance of this developmental pacing in our maturation and in our dying suggests that it is just as important in middle life and it is only a lack of perspective that keeps us from seeing it.

To illustrate their point, the critics of anti-aging medicine recall our moral intuitions about interrupting the development of a child at prepubescence. Biologist Leonard Hayflick writes:

The goal of slowing the aging process might be viewed in the same light in which we view slowing developmental processes. Slow physical or mental development in childhood is viewed universally as a serious pathology. If retarding the mental and physical development of someone from birth to age twenty years, for ten years in order to gain a decade of extra life is unattractive, then slowing one's aging processes in later life will not be attractive, and for the same reasons (Hayflick 2001–2002, p. 25).

What is worrisome about that scenario is not simply the psychological harm such a developmental distortion might produce. Nor is it just a matter of violating the child's rights to self-determination: those rights are not yet in full flower and it is the parents' role to protect, and to some extent define, their child's best interests. But some form of "autonomy," as self-governance, does seem violated here. The child's bodily development is no longer progressing on its own schedule, nor being driven by the complex, automatic interplay of genes and their reactions to the environment. This disruption of the child's "developmental autonomy," in turn, alienates his or her life story from the temporal narrative that characterizes our species. Postponing the normal biological changes of aging, the critics argue, constitutes a similar disruption. Whether or not the biological changes of aging are beneficial or harmful, they are meaningful: they and their natural timing constitute part of the normal life-cycle for human beings, and thus part of what it means to be human (Kass 2001). Intentionally distorting that cycle alienates the elderly from the definitive human life story, and dehumanizes them in the process. Adults should be taught to seek the meaning of the later stages of human development, and biomedical research should focus on making the experience of that part of life as healthy and pleasant as possible, without interfering in its essential rhythm (Callahan 2000). For those who see human life as a fixed natural cycle from the dependence of childhood to the dependence of old age, or for whom the discomforts of dying are as important as the discomforts of birthing for full human experience, even delaying age-associated morbidity beyond "premature aging" would produce an inappropriately distorted life experience.

In light of Hayflick's analogy to delayed maturation, it is instructive to think about the circumstance in which we do humor people's concerns about the harms of aging without taking them as serious calls to action. When my daughter was 12, she once responded to my description of her behavior as " adolescent" with hot denial, saying she was not a teenager and did not want to become a teenager, because teenagers were all "gross." She had already seen her sister suffer the bodily indignities of puberty, the aches and awkwardness of accelerated growth, the loss of previously enjoyed mutual interests, the runaway emotions, the truncation of sociability, and the decrease in energy that seemed to her to characterize that stage of life, and she saw all those changes as distinct harms. If she could have "compressed the morbidity" of adolescence by postponing those changes,

I am sure she would have. In fact, she probably was at her lifetime physical peak on most measures of human health and fitness; why not let her maximize the benefits and opportunities that level of health can provide?

"Growing pains" is what these outbursts of developmental anxiety are called at our house, and they tend to be humored indulgently as understandable, but childish. If the anxiety persisted, we might worry about "Peter Pan syndrome," and if it seemed driven by some social aspiration or pressure, like excelling in gymnastics, we would bristle protectively and rethink her activities. But we would not try to manipulate her endocrine system to prevent or postpone her adolescence, and most people would accuse any parents or physicians that did so with committing a grave moral wrong. Why is that and what does it imply for the ethics of adult anti-aging interventions?

First, we believe that some of the changes she perceives as harms—the emergence of sexual maturity and the shifts in interests and activities—are not harms at all. We see them as just natural parts of growing up, which, on balance, offer more benefits than losses. Moreover, we expect most of the worst symptoms to be transient: complexions clear, emotions stabilize, and awkwardness dissipates as adolescents adapt to their new life phase, early adulthood. Postponing puberty would be to deny children the wider benefits of growing up, which are thought by most who have enjoyed them to be well worth the price of adolescence.

Critics of anti-aging medicine suggest that a similar argument might be made for the biological changes of late adulthood, if our society were not so pervasively influenced by the perspective of those who have not yet undergone them (Callahan 1993). That is, perhaps not all the changes that young adults count as the harmful losses of aging are harms at all, when seen from the other side. One familiar example of this is menopause: a loss of reproductive capacity, fraught with physical and emotional turbulence, but one which many women come to celebrate as opening new opportunities and life-pleasures (Martin 1985; Logothetis 1993). Similarly, in many societies the loss of physical strength and endurance that comes with aging is what allows one to relinquish responsibility for the labor of survival, and move into an even more important role as an "elder" for one's community (Moody 1986). Traditionally, even the frank health challenges of aging—failing senses, vulnerability to disease and accident—have been seen as contributing to the life experiences of the elderly in a way that gives them a level of equanimity and insight difficult to achieve at earlier stages in life (Post 2000). Amongst 20th century authors, the psychologist Erik Erikson looked to old age as a crucial source of generativity in the human life-cycle (Erikson 1982), and philosophers Daniel Callahan and Leon Kass argue that growing old provides special opportunities for teaching, wisdom, and altruism (Callahan 1993, 2000; Kass 1983, 1985, 2001). This does not mean we should be unconcerned about the major diseases that threaten human health in late adulthood, any more than we should become complacent about the increased risk of fatal traffic accidents in adolescence. But it does hint, critics argue, that attempting to intervene in the aging process itself, for all its attendant complaints, may be shortsighted and harmful, by denying adults the wider human experience of growing old.

Of course, arguing that the traditional human life-cycle is normative for human beings requires a good bit of philosophical work if it is not to be accused of making a virtue of necessity. Just because human beings have always lived their lives within a particular pattern of life experiences is not necessarily a reason to continue doing so. In fact, the social, technological, and biological dimensions of the "typical human life story" have been rewritten continuously over our species' history without diminishing the moral status of those people whose lives are made possible by that evolution. Given our history of pushing back the "natural" limits of our lives through technology, the burden of proof in this case, the advocates argue, is on the "naturalists" to complete their philosophical project convincingly (Overall 2003).

How prescriptively should medicine take the contemporary biological parameters of human life? After all, the physical burdens that accompany aging are more serious than those of adolescence and less transient. It is harder to adapt to life with fragile skin and brittle bones than to life with oily skin and lengthening bones. Moreover, we do not now live in a society designed to optimize the role of the elderly. Given the social realities of aging in our culture, many adults would consider the price of the late stages of human development high enough to warrant attempts to postpone and compress them as much as possible. Moreover, unlike pre-adolescents, middle-aged adults have seen enough of life to allow them to project themselves and their interests beyond their current age and appreciate the trade-offs involved in postponing aging. Advocates of anti-aging research can point out that respecting that human ability to project and pursue a life-plan is at the heart of what it means to respect self-determination and personal autonomy. In the bright light of that value within our society, how could responding to informed adult requests for anti-aging interventions with new research and (elective) clinical services be wrong?

Until it is clear why, in the light of all its other intrinsic values, it is important for medicine to conserve the human species in its current form, a commitment to "life-cycle traditionalism" in medicine can only count as a nidiosyncratic ideology, which autonomous physicians (and their patients) in a free society should have the right to assess, adopt, or reject as they will. While the "dehumanization" argument bears more development, at this point it cannot provide a strong basis for discouraging physicians from the practice of anti-aging medicine under the banner of the "compression of morbidity."

LOYALTIES: PROTECTING THE HUMAN SPECIES

One popular way to delimit the proper sphere of biomedicine is to draw a line against interventions that might take their recipients out of our biological species altogether. This is usually framed as a minimal proposal to ban the clearest possible cases of immoral manipulation, on which everyone could agree. From proponents of "responsible genetics" (Council for Responsible Genetics 1993) to defenders of our "genetic patrimony" as the "common heritage of all humanity" (Knoppers 1991), to the "anti-post-humanist" life-cycle traditionalists (Kass), the prospect of "species altering experiments" which might "direct" human evolution is provoking resistance.

A particularly prominent example of this resistance is the recent appeal for a U.N. convention on genetic technologies framed in terms of "the preservation of the human species." As its advocates put it: "cloning and inheritable genetic alterations can be seen as crimes against humanity of a unique sort: they are techniques that can alter the essence of humanity itself (and thus threaten to change the foundation of human rights) by taking human evolution into our own hands and directing it toward the development of a new species sometimes termed the "post-human." Casting genetic interventions as "species-endangering" is a clever strategy for international science policy-making. It enables policy-makers to invoke concern over bioterrorism, human extinction, and ecological recklessness, private exploitation of public resources, species integrity, the wisdom of nature, and the rights of future generations. However, as I have argued in more detail elsewhere (Juengst 2006), it does suffer from three important problems.

First, species are not static collections of organisms that can be "preserved" against change like a can of fruit; they wax and wane with every birth and death and their genetic complexions shift across time and space (Robert & Bayliss 2003). In our case, almost everything we do as humans affects that process. To argue, as some Europeans have, that everyone has the right to inherit "an untampered genome" only makes sense if we are willing to take a snap-shot of the human gene pool at some given instant, and reify it as the sacred "genetic patrimony of humankind"—which some come close to doing (cf. Mauron & Thevoz 1991). Of course, if it is only those alterations that literally produce new biological species that are of concern, as the authors sometimes suggest, the vast bulk of imaginable germ-line genetic modifications—merely changing alleles within the existing human genome—would not fall within the jurisdiction of this approach.

Second, there is a risk here of confusing the biological sense of "human" as an taxonomic term (like "canine" or "simian") and the word's use in "human rights," where it serves as a synonym for "natural," "inalienable," or "fundamental" to distinguish that class of moral claims from other conferred, negotiated, or legislated rights. Obviously it is not enough to be taxonomically human to enjoy human rights: human tissue cultures and human cadavers show us that. Is it even necessary to be taxonomically human to enjoy human rights? There are many candidates for the qualities that serve to give us our inalienable rights (the draft Convention singles out reason and conscience), but none hinge on a taxonomic designation. So why is it that even dramatically speciating interventions "might cause the affected children to be deprived of their human rights"?

Finally, and perhaps most importantly, this approach simply risks distracting people from the fact that for the foreseeable future our most pressing human rights challenges from genetics will be *intra-* not interspecific. As Alex Capron pointed out two decades ago, the threat of genetics is not the post-human "Other" (Capron 1990). Rather, it is the risk that genetic information will be used to hammer scientific wedges into the social cracks that we have always used to make "others" out of each other. We should eschew any proposals to use genetic interventions—or, for that matter, genetic screening techniques—as public health tools to reduce the incidence of genotypes associated with genetic disease and disability in the population, as if particular genotypes represented a form of expensive "pollution" that could and should be cleansed from the gene pool. This way of thinking reduces the identities of people to their genotypes, and undermines our commitment to the moral equality of people despite their biological diversity.

UNNATURAL ABILITIES, NATURAL TALENTS, AND CHAMPIONS IN SPORT

Finally, a third important invocation of the natural lies at the heart of some of the most sophisticated of today's policy discussions about the use of performance-enhancing interventions in elite sports. This discussion is interesting because of the way it illuminates the flip side of appeals to human nature: i.e., what we mean when we accuse a particular biomedical intervention of having "unnatural" effects.

In 2006, the World Anti-Doping Association (WADA), which polices performance-enhancing drug use in elite sport, added "gene doping" to their list of prohibited activities (Hall et al. 2006), in an effort to anticipate the hypothetical use of human gene transfer techniques to improve athletic performance. Although no human gene transfer protocols that might enhance athletic performance yet exist, the insertion of normal human genes into the muscles of athletes would stimulate the production of natural proteins that would be very hard to detect, making the prospect especially troubling for those concerned with the ethics of sport.

The ethical questions that "gene doping" most commonly prompt (beyond safety concerns) are questions about the moral quality of an athlete's achievements: questions about fairness, natural achievement, and the spirit of sport as a social practice. Given the expense and technical complexity of current gene doping scenarios, the fairness questions are probably the most pressing in practice. But the most powerful critiques of gene doping in principle flow from the latter questions, and it is in that context that fundamental sports values at stake emerge—for better or worse.

At one level, gene doping would seem to be simply a problem of cheating. If the rules of sport forbid the use of genetic performance enhancements, then their use becomes illicit and confers an unfair advantage against athletes who either accept the rules of the game or do not have access to the enhancement interventions. That advantage, in turn, can create pressure for more athletes to cheat in the same way, undermining the basis for the competitions at stake and exacerbating the gap between athletes who can afford gene doping and those who cannot (Murray 1983). But of course, this begs the question of why gene doping should be against the rules of sport in the first place. As long as gene doping interventions can neither be distributed fairly nor reconciled as part of the inherent advantages that "come with the territory" for some fortunate elite athletes, then an argument from fairness can support the current WADA gene doping ban. But for those who can afford it, what would be ethically suspect about a mirror-image of the "Special Olympics" for athletes with disabilities: a "SuperOlympics," featuring athletes universally equipped with the latest modifications and enhancements (Munthe 2000)? For answers to that challenge, critics of gene doping must look to the other questions that "enhancement" flags: questions about the naturalness of athletic achievements and the meaning of sport.

To explain why the rules of sport should not be rewritten to legitimize performance-enhancing interventions like gene doping, most critics turn at this point from the consequences of gene doping to its intrinsic merits as a means for improving athletic achievements. Athletic achievements, they argue, are praiseworthy only to the extent that they are accomplished "naturally" by the athletes themselves, both because those are the achievements for which athletes can legitimately claim personal responsibility and credit, and because only those achievements display the distinctly human excellences that sport celebrates (Hoberman 1992). Some methods of performance enhancement—training, nutrition, better equipment—are compatible with natural athletic achievements, and are therefore unproblematic on this view. Other methods, however, like gene doping, are unnatural shortcuts to achievement, producing gains the athlete has not earned and yielding hollow achievements that only cheapen the human excellences they are supposed to celebrate (cf. President's Council 2003; Schnieder and Butcher 2000).

As Hoberman points out, this focus on the moral merits of the means athletes use to improve their performance has meant that "the distinction between what is 'natural' and what is 'unnatural' is at the heart of the twentieth-century controversy over the use of performance-enhancing drugs in sport" (Hoberman 1992, p. 104). But how should this distinction be understood? As I have argued elsewhere, appeals to the "unnaturalness" of biomedical technologies are relatively common within bioethics, but are used to signal at least three different kinds of moral concerns. Either they are concerns about the blurring of natural boundaries, or "abomination," concerns about transgressing on the supernatural, or "blasphemy," or concerns about undermining authenticity, or "fraud" (Juengst 2009). Invocations of unnaturalness in sports ethics focus on the third set of concerns, and, in their most sophisticated versions, are explicated against a vision of "the spirit of sport" that makes the celebration of "natural talents" the main concern of the enterprise.

In the rhetoric of sports ethics, problematic interventions are often labeled in ways that highlight the distinction between the natural and the artificial, without much analysis. Thus, we are warned against "synthetic steroids" (Hoberman 1992), "designer doping" (Joyner 2004) "the scientific manufacture of athletes" (Tannsjo 2000), "artificially-induced hypoxic conditions" (WADA 2006), and even "artificial genes"(!) (Unal & Unal 2004), but we are seldom told precisely why these adjectives matter.

The distinction between the natural and the artificial is common in other bioethical debates as well, over issues ranging from the right to a "natural death" to the moral merits of "artificial reproductive technologies." Interestingly, however, it is usually invoked when the processes in question—death, conception, respiration, hydration, aging—are basic biological processes we share with other living organisms. The distinction's point is to draw moral attention to biomedical attempts to "artificially" replicate, control, or improve upon these processes, by emphasizing that these "natural" processes are not human inventions in the first place. As a call to arms in sport, unfortunately, this contrast falls flat in the face of the fact that sport itself is, in the first place, a human invention, practiced with artificial rules and dependent on multiple technologies of human manufacture. This leads some to dismiss this interpretation quickly as incoherent (cf. Mehlman 2003, pp. 95–101); perhaps too quickly, since in fact it can point to the most plausible set of principled reasons to be concerned about "enhancement" in the sports context.

Some of those concerned to flesh out the moral meaning behind this distinction argue that explaining the unnaturalness of gene doping in terms of its artificiality is to express the concern that the achievements it would make possible would be not be achievements authored by the athlete, for which he or she could legitimately take credit. Under this view, gene doped victories would be fake, hollow, or inauthentic because they literally would not be the athlete's victories (cf. President's Council 2003, pp. 124–134). Even the WADA Ethics Committee takes this approach when it argues that, for example:

> Our analysis of artificially induced hypoxic conditions to modify performance alerted us to an important distinction: between technologies that operate on the athlete and in relation to which the athlete is a merely passive recipient, versus technologies with which the athlete actively engages and interacts as part of the process of training and competing in order to enhance performance. From the individual athlete's point of view, my responsibility for my performance is diminished by technologies that operate upon me, independent of any effort on my part. . . . The human athlete utilizes, masters and controls the technology, not the other way round (WADA 2006).

The problem with this interpretation, of course, is the difficulty in disentangling the many conscious decisions athletes make that are accepted as legitimate, even praise-worthy contributors to their responsibility for their accomplishments in the normal course of sport from a decision to undertake a gene transfer intervention. If accepting a particular diet can be a legitimate feature of an athlete's "victory narrative," why not the decision to accept gene doping? A gene doped victory may not be morally praiseworthy, but it would still be authored by the athlete. As Cole-Turner writes,

> The fact that I write at a computer makes writing easier by eliminating retyping and other frustrations, but writing itself is an intense struggle and it will remain so under any technological condition. The technological advance does not eliminate the struggle so much as relocate it; indeed, it makes it possible to eliminate secondary aggravations and focus attention on the core struggle at hand, namely, expressing new insights in just the right words. Even if technology increased our cognitive ability itself . . . so that we could calculate or write or think more clearly, these activities would still be a struggle in the face of even greater intellectual challenges to which we human beings inevitably set ourselves (Cole-Turner 2008, p. 156).

Claudio Tamburrini goes even further to suggest that gene doping may even allow us to focus more explicitly on the features of athletic performance that do deserve our praise and admiration:

> We will no longer admire sport stars for their outstanding physical traits (they will be genetically designed). Rather, we will continue to admire them for all the sacrifices endured to actualize their genetic predisposition. In the same way that today's natural talents cannot do without hard training, the genetically transformed athlete will have to devote herself to her discipline in a goal-directed and professional manner . . . Our admiration for sport heroes will, to a much higher degree than today, concentrate on their dedication and efforts, rather than on a fortuitous physiological predisposition. I cannot see anything unsound in such a development" (Tamburrini 2002, p. 266).

Of course, unless the naturally talented are somehow prohibited from using gene doping (returning us to today's situation), the availability of such performance enhancements will never have this leveling effect. But even if it could, others can see, *contra* Tamburrini, something unsound in such a development. What makes a genetically engineered sport hollow, they argue, is not that it renders individual athletes frauds, but rather that it simply misses the point of sport as a social practice (Murray 1987).

On this view, even if the explicit rules of a sport are changed to allow new enhancements, their use would erode the value of the achievements they make possible, by undermining the practices that make the achievements valuable in the first place. These practices (disciplined training, graceful play, good teamwork, etc.) add value to the achievements because they are understood to be admirable social practices in themselves. Wherever a biomedical intervention is used to bypass an admirable social practice, then, the improvement's social value is weakened accordingly (Murray 1987; Schnieder & Butcher 2000; Loland 2002).

Interpreting enhancement interventions as those that short-circuit admirable human practices has special utility for sports ethics. To the extent that biomedical shortcuts allow specific accomplishments to be divorced from the admirable practices they were designed to signal, the social value of those accomplishments will be undermined. This line of thinking is becoming increasingly influential in sports ethics, and seems to offer a stable moral platform from which to evaluate the gene doping issue and the fundamental sports policy choice it raises. It means that for institutions interested in continuing to foster the social values for which they have traditionally been the guardians, choices will have to be made. Either they must redesign the game to find new ways to evaluate excellence in the admirable practices that are not affected by available enhancements, or they must prohibit the use of the enhancing shortcuts. However, knowing which way to go means having at least the sketch of theory of sport as a social practice and of the values that animate it. Fortunately, the WADA Ethics Committee itself offers such a sketch, which is instructive to examine.

If in fact the problem with gene doping is that it misses the point of sport, what intrinsic values seem to be at stake? In a recent statement on hypoxic training chambers, WADA, echoing Sigmund Loland, Tom Murray, and others, says that the spirit of sport lies in "the celebration of natural talents and their virtuous perfection," which they explain in the following way:

> What is it that people around the world find honorable, admirable and beautiful about sport? In one important sense, sport is a celebration of human variation. Not all of us have the physiology or anatomy to be a great swimmer, hurdler, skier or discus thrower. Yet, biology is not all that matters: the most naturally gifted athletes must work to perfect their talents. . . . The spirit of sport, as we understand it, celebrates natural talents and their virtuous perfection. We say "virtuous" in this context because virtues are qualities of character admirable in themselves, the qualities that outstanding athletes develop and embody in their quest for excellent performance. Some means we respect and want athletes to employ exemplify aspects of character that we admire in people more generally, such as fortitude, dedication, self-discipline, the willingness to suffer in the service of a worthy cause, courage, and strategic wisdom. . . . So, for any particular means for enhancing performance . . . the crucial test will be whether it supports or detracts from sport as the expression of natural talents and their virtuous perfection (WADA, 2006).

This sketch has several important features that bear scrutiny. First and foremost, sport celebrates human variation: that is, the differences between individual human bodies. But not all human differences are of interest in sport. Sport is concerned with celebrating differences in natural talents and the virtues that can be displayed in attempts to differentiate one's own talents even further. The virtues that sport celebrates are socially admirable habits and traits in and of themselves, and their promotion is what gives sport is social value as a practice. However, within the practice, the virtues are instrumental (as either side constraints or facilitators) to the "perfection" of the athlete's natural talents—i.e., to their differentiation from other talents. Moreover, not all natural human talents are relevant in sport. Sport, we learn, is concerned with celebrating the variation in human natural talents that leads athletes to their goal of "excellent performance."

What constitutes excellent athletic performance? WADA's opening question offers three scales for judgment: a performance's moral virtues ("honorableness"), its aesthetic qualities ("beauty") and its place in a hierarchy of human accomplishments ("admiration"). Honorableness, as WADA suggests, is a necessary condition for performance excellence, and to a lesser extent beauty may be as well. But neither seem sufficient conditions, alone or together. As other philosophers of sport have pointed out before, an athletic performance can be beautiful and well-played, but if it does not win, either by beating the athlete's own record or the competition, it is not as excellent as an equally beautiful, well-played event that does produce a champion. Performance excellence in sport, in other words, seems to be inevitably and intrinsically about the comparative ranking of performances. Thus, Sigmund Loland concludes:

> In spite of great diversity in sport-specific goals, then, it is possible to formulate a general goal that characterizes sports competitions as such: the goal of sports competitions is to measure, compare and rank two or more competitors according to athletic performance. This goal seems to be common to all sports however diverse their ethos. It defines sport's characteristic social structure and I shall therefore call it the structural goal of sport competition (Loland 2002, p. 10).

The key role of hierarchical ranking in sport is often ignored in the official rhetoric of sports organizations, but as sports ethicists like Coakley acknowledge, fixation with hierarchical ranking—with competition, contest, score-keeping, record-breaking, championship, victory, and defeat—is pervasive in the everyday practice of sport (Coakley 1998). In ordinary discourse, sports remains true to its historical roots in the war games of ancient and medieval "arenas," "tournaments," and "(battle)fields." As Tannsjo has provocatively pointed out, the admiration of champions can easily become contempt for the losers: to "beat," to "own," to "dominate" the competition is to metaphorically subjugate them, just as conquering champions have always enslaved those they overcome (Tannsjo 2000). While the hierarchies established by sport are usually more benign than the language of sports (and Tannsjo, for that matter) suggests, the aptness of our bellicose metaphors shows a similar human urge at work: the urge to establish interpersonal and intergroup hierarchies and locate ourselves within them.

Perhaps most ordinary sports metaphors also miss the point of sport and betray its true spirit. But the value of comparative ranking also insinuates itself into careful attempts to reason about ethical issues in sport like gene doping, as our tour of the debate has shown, and that is even more telling. Returning to WADA's description of the spirit of sport, perhaps one could say that the many different forms of sport accommodate the diversity of human body types in an appreciative and equitable way. But within any one sport, the "celebration of human variation" is always comparative and hierarchical. The spirit of sport is the celebration of the differences in the human talents and virtues that allow athletes to honorably and beautifully accomplish achievements that place them higher in a hierarchy of human excellence than their competition.

So far, nothing about this spirit of sport would seem to be betrayed by the prospect of safe and openly available gene doping. As the apologists for gene doping insist, genetic interventions would not diminish the need for effort, courage, or self-discipline in sport, because the bar to victory is simply lifted higher to require them (Miah 2004; Tamburrini 2002). In fact, genetic interventions could help produce even greater champions to celebrate by elevating the starting points for their accomplishments: the physical and mental talents which, as athletes, they set about to virtuously perfect.

But there is still another feature of WADA's description that has not been examined: the fact that in its invocation of the spirit of sport, WADA's ethics committee restricts the kinds of talents sports celebrates to "*natural* talents." The term is being used here to qualify the kinds of talents which athletes attempt to virtuously perfect and which authentically excellent (honorable, beautiful, and admirable) athletic performances celebrate. We have already put aside the possibilities that this use of "natural" might be intended to exclude monstrous or blasphemous talents, or talents aided by technological artifice *per se*, for the same reasons that brought us to this point in the analysis. Only one other plausible interpretation of "natural" in this phrase presents itself: "natural talents" are the talents you were born with—i.e., inherited characteristics over which athletes have no control and which are ultimately traceable to particular combinations of ancestors and their genes. As Sigmund Loland concludes:

> Now I can say more precisely what I mean by talent in sport. It is an individual's genetic predisposition to develop phenotypes of relevance to performance in the sport in question. The distribution of talent in the natural lottery is a random process (Loland 2000, p. 69).

On WADA's interpretation, in other words, the spirit of sport is in part a celebration of differential genetic endowments, distributed through the "natural lottery" of genealogy. Sport creates a system of values, virtues, and practices designed to hierarchically grade people in terms of their (virtuously perfected) inherited traits and glorify the best specimens as champions. This is a turning point for our analysis: biomedical interventions at the genetic level would miss the point of the sport if this view is correct, because gene doping would undermine the ability of sport to distinquish those who passively inherited their talents from their progenitors from those who actively acquired them from their physicians. As Sandel says:

> Some might say effort: the problem with drugs is that they provide a shortcut, a way to win without striving. But striving is not the point of sports; excellence is. And excellence consists at least partly in the display of natural talents and gifts that are no doing of the athlete who possesses them. . . . No one believes that a mediocre basketball player who works and trains even harder than Michael Jordan deserves greater acclaim. The real problem with genetically altered athletes is that they corrupt athletic competition as a human activity that honors the cultivation and display of natural talents" (Sandel 2004, p. 55).

In other words, gene doping is wrong for athletes to pursue and sports medicine to provide because it compromises the ability of sport to segregate and elevate genetically advantaged athletes from their disadvantaged competitors, which is a key element of the spirit of sport and one of the intrinsic values of the enterprise. This will be true even if gene doping is proven safe and effective, and even if it could be provided equitably to every competitor in non-coercive ways.

After all our conceptual orienteering, why is reaching this destination so disturbing? Perhaps it is because of the uncomfortable light that the problem of gene doping throws back on the spirit of sport itself and the broader questions it raises for sports ethics. The very prospect of gene doping forces sport to notice that the "natural lottery" of genetics is getting less "random" all the time as our knowledge of human genetics grows, and as that happens the moral justification for celebrating human genetic hierarchies as facts of nature begins to falter. It is one thing to acknowledge, as perhaps unfortunate but not unfair, that human talents vary across individuals in ways that allow some to do some things that others cannot. To admire those inborn differences as beautiful and to reward those who seek to (virtuously) perpetuate and increase them seems arbitrary, but no less pernicious than other idiosyncrasies of social taste, like celebrity worship. But to glorify those genetic disparities to the extent of prohibiting their abatement when biomedicine provides the ability to do so exposes a fundamental tension with the spirit of sport itself: while the variable genetic distribution of human talent may be a natural fact, the creation of social hierarchies out of that distribution is a distinctly human project, and never a random one. For sports ethicists, this raises a challenging question for further discussion: as a human project, what differentiates the hierarchical ranking of genetic talents by sports institutions from the reciprocal devaluing of genetic defects by insurance risk underwriters? In both cases, the internal logic of the practice is clear, but their use of inherited genetic identity as a social ranking criterion raises large questions about the fairness of the enterprise as a whole.

This may be the reason for the apocalyptic tone with which sports authorities frame the issue of gene doping. After all, in most discussions of the ethics of genetic enhancement outside the sports context, a key concern is to avoid the creation of a "genobility" and the errors of genetic essentialism and risks of genetic discrimination that would entail (Mehlman 2003). Even within medicine, the strongest case against going "beyond therapy" with genetic interventions is the danger of exacerbating invidious social divisions by reducing them to individual genetic advantages and disadvantages. In a world struggling to realize a commitment to treating people as moral equals despite their biological differences, a social practice that creates and glorifies hierarchies of genetic endowment seems anachronistic and slightly ominous. As Sandel acknowledges about his defense of the role of sport's commitment to genetic stratification:

> This is an uncomfortable fact for democratic societies. We want to believe that success, in sports and in life, is something we earn, not something we inherit. Natural gifts, and the admiration they inspire, embarrass the meritocratic faith; they cast doubt on the conviction that praise and reward flow from effort alone (Sandel 2004, p. 55).

To discover that the principal problem with gene doping is that it threatens sport's commitment to the creation and promotion of human genetic stratification should give us pause. On one hand, of the many ways humans use inherited traits to create interpersonal hierarchies, athletic competition is among the most benign. When it is "just a game," comparative interpersonal ranking in terms of genetic identity in sports is a welcome substitute for blood feuds, racism, and genocide. What the gene doping debate suggests, however, is that sport may have more in common with these extreme human vices than we would like to think. Being one's peers' last choice as a teammate hurts enough in a friendly pick-up game. But when sport becomes a matter of national pride and a source of economic opportunity, athletic losers risk more than simply admiration and social status; like insurance applicants with genetic susceptibilities, less naturally talented athletes risk access to important social benefits and potential life plans. In this regard, the apocalyptic challenge that genetic engineering poses to sport reduces to a simple question that echoes Cole-Turner's concern to "relocate the struggle" of human achievement on what really matters: are there ways to enjoy, appreciate and even show off our bodies and abilities without requiring someone else to lose social standing on genetic grounds?

CONCLUSION

It is important to listen when claims are made that some form of biomedicine violates human nature, even in public discussions of policy within a pluralistic society. Whether the concern is the distortion of some constant of the human condition, like senescence, or a "species altering" threat to our collective gene pool, or the corruption of practices designed to celebrate the inherited human traits we value most, these appeals all signal that the intervention in question has deep implications for who we want to be, given who we have been. For people living in community, these are fundamentally questions of politics and policy. In an era when the inevitability of continued human evolution is assumed, respecting what we have (or have not) inherited from our parents does not fulfill the need to decide which promises we would like to make to our children. Invocations of particular legacies, loyalties, or rankings from the past can provide fodder for arguing over the positive visions of human nature that should guide those promises. But in communities that accept the possibility of a pluralism of promises, such invocations should also trigger another policy-making response: the need for policy-makers to protect the interests of those excluded from their visions, even while we discuss their merits. The natural human gene pool has no top, bottom, edges, or direction; it cannot be "used up," "diverted," "purified," or "polluted." The reservoir of human mutual respect, good will, and tolerance for difference, however, seems perennially in danger of running dry. That is our human nature's truly fragile heritage that we should seek to preserve in monitoring biomedical research on behalf of the future.

REFERENCES

Annas, G., Andrews, L. & Isasi, R. (2002). Protecting the endangered human: Toward an international treaty prohibiting cloning and inheritable alterations. *American Journal of Law and Medicine* 28: 151–178.

Bernstein, J., Perlis, C. & Bartolozzi, A. (2000). Ethics in sports medicine. *Clinical Orthopaedics and Related Research* 378: 50–60.

Buchanan, A., Brock, D., Daniels, N. & Wikler, D. (2000). *From chance to choice: Genetics and justice.* New York: Cambridge University Press.

Burke, M. & Roberts, T. Drugs in sport: An issue of morality or sentimentality? *Journal of the Philosophy of Sport* 24: 99–113.

Callahan, D. (1993). *The troubled dream of life: Living with mortality.* New York: Simon and Schuster.

Callahan, D. (1995). Aging and the life cycle: A moral norm? In D. Callahan, E. Topinkova, & R. H. Meulen. (Eds.), *A world growing old: The coming health care challenges.* Washington D.C.: Georgetown University Press. 21–27.

Capron, A. (1990). Which ills to bear? Reevaluating the "threat" of modern genetics. *Emory Law Journal* 39 (Summer): 665–696.

Coakley, J. J. (1998). *Sport in society: Issues and controversies* (6th ed.). Boston: McGRaw-Hill.

Cole-Turner, R. (1998). Do means matter? In E. Parens (Ed.), *Enhancing human traits: Ethical and social implications*. Washington D.C.: Georgetown University Press. 151–161.

Council for Responsible Genetics (1993). Position statement on human germ-line manipulation. *Human Gene Therapy* 4: 35–39.

Daniels, N. (1986). *Just health care*. Cambridge: Cambridge University Press.

Daniels, N. (1992). Growth hormone therapy for short stature: Can we support the treatment/enhancement distinction? *Growth, Genes and Hormones* 8 (Suppl. 1): 46–48.

Erikson, E. (1982). *The life cycle completed: A review*. New York: W. W. Norton.

Forbes, W. F. & Thompson M. E. (1990). Age-related diseases and normal aging: the nature of the relationship. *Journal of Clinical Epidemiology* 43(2): 191–193.

Fries J. F. (1980). Aging, natural death, and the compression of morbidity. *New England Journal of Medicine* 303: 130–136.

Good, B. (1994). *Medicine, rationality, and experience: An anthropological perspective*. New York: Cambridge University Press.

Hall, J., Dunlap, J., Friedman, T. & van Hyningen, V. (2006). Gene transfer in sports: An opening scenario for genetic enhancement of normal human traits. *Advances in Genetics* 51: 37–49.

Hayflick, L. (2001–2002). Anti-aging medicine: Hype, hope and reality. *Generations* 24(4): 20–27.

Hoberman, J. (1992). *Mortal engines: The science of performance and the dehumanization of sport*. Caldwell, NJ: The Blackburn Press.

Joyner, M. (2004). Designer Doping. *Exercise and Sport Science Reviews* 200: 81–82.

Juengst, E. (2006). Alter-ing the human species? Misplaced essentialism in science policy. In J. Rasko, G. O'Sullivan & R. Ankeny (Eds.), *The ethics of inheritable genetic modification*. Cambridge: Cambridge University Press: 149–159.

Juengst, E. (2009). Annotating the moral map of enhancement: gene doping, the limits of medicine and the spirit of sport. In T. Murray (Ed.), *Ethics, genetics and the future of sport: Implications of genetic modification and genetic selection*. Washington D.C.: Georgetown University Press. 175–204.

Juengst, E., Binstock, R. Mehlman, M., Post, S., & Whitehouse, P. (2003). Biogerontology, "anti-aging medicine," and the challenges of human enhancement. *The Hastings Center Report* 33 (July/August): 2–10

Kass, L. (1983). The case for mortality. *The American Scholar* 52: 173–191.

Kass, L. (1985). *Toward a more natural science: Biology and human affairs*. New York: Free Press.

Kass, L. (2001). L'Chaim and its limits: Why not immortality? *First Things* 113(May): 17–25.

Ljungqvist, A. (2005). The international anti-doping policy and its implementation. In C. Tamburrini & T. Tannsjo. *Genetic technology and sport: Ethical questions*. New York: Routledge. 13–19.

Logothetis, M. L. (1993). Disease or development: Women's perceptions of menopause. In J. Callahan (Ed.), *Menopause: A midlife passage*. Bloomington, IN: Indiana University Press.

Loland, S. (2002). *Fair play in sport: A moral norm system*. New York: Routledge.

Martin, M. (1985). Malady and menopause. *Journal of Medicine and Philosophy* 10: 329–339.

Mauron, A. & Thevoz, J.-M., (1991). Germ-line engineering: A few European voices. *Journal of Medicine and Philosophy* 16: 649–666.

Mehlman, M. (1999). How will we regulate genetic enhancement? *Wake Forest Law Review* 34: 671–714.

Miah, A. (2004). *Genetically modified athletes: Biomedical ethics, gene doping and sport.* New York: Routledge.

Munthe, C. (2000). Selected champions: Making winners in the age of genetic technology. In T. Tannsjo & C. Tamburrine (Eds.), *Values in sport: Elitism, nationalism gender equality and the scientific manufacture of winners.* New York: Routledge. 217–221.

Murray, T. (1983). Drugs, sports, and ethics. In T. Murray, W. Gaylin & R. Macklin (Eds.), *Feeling good and doing better: Ethics and nontherapeutic drug use.* Clifton, NJ: Humana Press. 107–126.

Murray, T, (1987). The ethics of drugs in sport. In Richard Strauss (Ed.), *Drugs and performance in sports.* Philadelphia: W. B. Saunders. 11–21.

National Institute on Aging (2001). Action plan for aging research: Strategic plan for fiscal years 2001–2005. Washington, D.C.: U.S. Department of Health and Human Services.

Olshansky, S. J., Hayflick, L. & Carnes, B. (2002). Position statement of human aging. *Journal of Gerontology: Biological Sciences* 57A; 8: B292–B297.

Overall, C. (2003). *Aging, death and human longevity: A philosophical inquiry.* Berkeley, CA: University of California Press.

Parens, E. (Ed.) (1998). *Enhancing human traits: Ethical and social implications.* Washington, D.C.: Georgetown University Press.

President's Council on Bioethics (2003). *Beyond therapy: Biotechnology and the pursuit of happiness.* Available from: <http://bioethics.georgetown.edu/pcbe/reports/past-commissions>. Accessed 14 June 2011.

Robert, J. S. & Baylis, F. (2003) Crossing species boundaries. *The American Journal of Bioethics* 3(3): 1–14.

Sandel, M. (2004). The case against perfection: What's wrong with designer children, bionic athletes and genetic engineering. *Atlantic Monthly* (293): 51–62.

Schnieder, A. & Butcher, R. B. (2000). A philosophical overview of the arguments on banning doping in sport. In T. Tannsjo & C. Tamburrini (Eds.), *Values in sport: Elitism, nationalism, gender equality and the scientific manufacture of winners.* New York: Routledge. 185–199.

Scully, J. L. & Rehmann-Sutter, C. (2001). When norms normalize: the case of genetic "enhancement." *Human Gene Therapy* 12: 87–95.

Simon, R. L. (1991). *Fair play: Sport, values and society.* Boulder, CO: Westview Press.

Stout, J. (1988). *Ethics after Babel: The languages of morals and their discontents.* Princeton: Princeton University Press.

Tamburrini, C. (2002). After doping, what? The morality of the genetic engineering of athletes. *Sport technology: History, philosophy and policy* 21: 243–268.

Tannsjo, T. (2005). Genetic engineering and elitism in sport. In C. Tamburrini & T. Tannsjo (Eds.), *Genetic technology and sport: Ethical questions.* New York: Routledge. 57–70.

Tannsjo, T. (2000). Is our admiration for sports heroes fascistoid? In T. Tannsjo & C. Tamburrine (Eds.), *Values in sport: elitism, nationalism gender equality and the scientific manufacture of winners.* New York: Routledge. 24–38.

Unal, M. & Unal, D. (2004). Gene doping in sports. *Sports Medicine* 34: 357–362.

World Anti-Doping Agency (2005). Available online at http/www.wada-ama.org/rtecontent/document/list_2005.pdf.

World Anti-Doping Agency (2006). *WADA Note on artificially induced hypoxic conditions.* Available from <http/www.wada-ama.org/rtecontent/document> Accessed May 24, 2006.

6 In the Stars or in Our Genes
The Languages of Fate and Moral Responsibility
Larry R. Churchill

INTRODUCTION

The title of this essay was taken from James Watson's quip to a *Time* magazine reporter that we used to believe that our destiny was in the stars, but we now know it is in our genes (Nash & Thompson 1989). Notice the contrast not just between metaphysical and biological explanations, but also the contrast between believing and knowing. His remark captures nicely the way in which the Human Genome Project (HGP) has enlivened and legitimated material explanations of the human condition, and leads readily to deterministic idioms. Determinism has become our default framework for understanding genetics.

This essay is about how determinism can influence notions of responsibility. My focus will be on some of the various languages of fate, often considered inimical to moral responsibility and thereby a barrier to ethical reflection and dialogue. Genetic determinism is a powerful and pervasive version of this way of thinking, deeply embedded in multiple communities in contemporary North American culture (McGee 1997; Kaplan 2000). For example, an off-hand sort of determinism is implied in the linguistic habits of scientists and physicians when they speak of "a gene for _____," even when they know the story is more complex. Determinism is also evident in media stories, and in the interpretive frameworks of the public, who often have no ready access to the more complex narratives of science and medicine, and may not know better (Hubbard & Wald 1993; Nelkin & Lindee 1995; Thompson 1994). I am also particularly concerned here with the differences between the deterministic language of genetic science and some of the idioms of religious fate. I will argue that "fate" and "destiny" are less dangerous metaphors in religious hands than in scientific and medical ones. Determinism in religious or spiritual life is less hazardous to ethics because it is often paradoxically coupled with high degrees of practical responsibility. In this sense, religious notions of fate or destiny often, though not always, probe to different and deeper levels of self-understanding, because they are not tone-deaf, as science often is, to mystery or intolerant of surface contradictions. Religiously/spiritually attuned people, however we might want to define these attributes (Churchill 2009), often understand that life has to be lived on many strata of conceptual understanding and practical engagement. It is this more stratified and complex understanding of the human situation that students of religion find fascinating, and that I believe are not sufficiently appreciated by those in positions of social and professional authority, such as scientists, clinicians, and bioethicists. I hope this essay will provide a glimpse into this complex set of more stratified, lived experiences in which various forms of determinism and authentic choice can coexist.

DETERMINISM AS A THREAT TO ETHICS

The central belief of genetic determinism is that we *are* our genes—that our actions, attitudes, characteristics, and health status are fixed in some fundamental and pervasive way by our genetic make-up. The paradigm case for genetic determinism is Huntington's disease, and the poster child for Huntington's is musician Woody Guthrie. Huntington's disease is a fatal neurodegenerative illness that typically first affects people in their thirties and forties, has no means of prevention or cure, and is linked to the mutation of a single gene on the tip of chromosome 4. While there is variation in terms of disease onset, individuals with the relevant mutation are fixed on an irreversible course of debilitation and early death, regardless of other aspects of their genetic complement, or differences in social or physical environments. Woody Guthrie died at age 55 following years of misdiagnoses and progressive deterioration.

Deterministic views of disease may seem more threatening to ethics when they are psychiatric or behavioral, and are characterized by loss of control, such as intermittent explosive disorder (IED). IED is characterized by extreme expressions of anger disproportionate to the context in which they occur, and accompanied by a sense of inability to resist them (American Psychiatric Association 2000). Episodes of IED sometimes result in "road rage," or domestic abuse, often involve personal or property damage, and are usually followed by embarrassment and remorse. Therapies involve learning to recognize symptoms and mastering techniques for impulse control. Although there are currently no precise genetic links to IED, the causes are routinely described as a combination of genetic and environmental factors.

Deterministic views of the human situation, whether they involve biological or behavioral disorders—or some combination—are often thought to strike at the heart of ethics because they seem to diminish or remove opportunities for choosing; the larger the range of determinism, the smaller the sphere for choice, such that a thorough-going determinism would undermine ethics altogether. Simon Blackburn says it well in *Being Good* (2001,

p. 44). "The threat is the paralyzing effect of realizing that we are what we are: large mammals, made in accordance with genetic instructions about which we can do nothing." The moral enterprise of arguing, sorting persuasive reasons from less convincing ones, and making choices based on one's value commitments is thereby rendered hopeless. A moral decision to take a different course than our genetic programming permitted would be like a prohibition against growing hair, or forbidding hunger or thirst, Blackburn claims. All are equally out of our control. Ethics becomes pointless, simply because our so-called "choices" are not really up to us at all. In short, no freedom of choice means no ethics, and hence no responsibility, no moral evaluations of agents or outcomes. At least that is a common assumption among moral philosophers. Protecting and enlarging the sphere of choice and confining responsibility to that sphere is a familiar theme in contemporary ethics, especially perhaps in the U.S. For example, Judith Jarvis Thomson (1971) has famously argued that a woman's responsibility for a fetus she is carrying begins when she chooses to continue the pregnancy. Hence, parenthood is a status one only need enter through the door of explicit consent. Whatever one's position on reproductive choices, Thomson's general point has wide resonance. If we think of moral responsibilities as contractual agreements between ourselves and others, voluntariness is clearly a central feature in determining what duties, if any, are owed.

VOLITIONAL AND STATUS RESPONSIBILITY

But surely this volitional aspect of responsibility is only part of the story. A quick review of the broad range of obligations most readers would readily acknowledge shows this notion of voluntary, willful responsibility to be too narrow and overly theoretical. Think here about responsibilities to and for family members—most of which we did not choose. For example, my duties to my aging father are ones I did not elect, except perhaps in terms of their extent. There is a minimal range of obligations for attending to his well-being that was simply given to me and my siblings. And while it would be fair to claim that I did in some sense *choose* to have children who came with lots of obligations, I did not choose *these* children, with their peculiar talents, needs, and aspirations. Perhaps it is more accurate to say that I did *intend* to have children, but could not have chosen these particular people as my children, since their presence in my life is a gift I am not wise enough to have chosen, even were it in my power. Readers will no doubt think of many other examples, both inside and outside family life.

John Silber has written elegantly on unchosen duties, which he terms "status responsibility" (1967, pp. 197–254). In a long and influential article Silber argues that "the fundamental basis of moral obligation is found in the nature and situation of the agent. These status elements are preconditions of the possibility of moral offense." He embraces a notion of responsibility characterized by a broad continuum of greater-to-lesser degrees of voluntariness, and finally argues that "human choice is not something isolated from the choosing person," and "at every instant, even in those acts of purest, freest and most voluntary choice, choice depends upon the being of the person and the matrix of his action." I begin with this notion of status responsibility to underline the fact that despite the appealing self-image of near-sovereignty over our responsibilities, and its equally appealing theoretical simplicity, voluntaristic concepts tell only part of the story, and often the lesser part of the story, of responsibility. The range of things for which we may appropriately find ourselves responsible is typically large; hence, efforts to lessen this load of duties may be quite appealing. While it may be theoretically troubling for the enterprise of ethics, determinism can also serve as a psychological safe harbor from an endless sea of obligations. Of course, we might limit responsibilities by accrediting only the obligations we have freely chosen, but it seems evident that this strategy defies common experiences. Determinism seems more attractive, especially if it has the backing of science.

THE DOGMA OF DETERMINISM

If lessening the load of responsibilities is the aim, there is hardly a better way to do so than by asserting that things are pretty much decided for us, that we are destined or fated to be who we are, do what we do, and that life has been largely designed for us. "That's just who I am," we sometimes say, suggesting a lessening of responsibility based on some presumably unchanging feature of our history or personality. This is precisely the sort of comfort that genetic determinism provides. And while determinism may be problematic for philosophers like Blackburn, to non-theorists it can be a kind of relief.

I will return to this set of problems later, but for now I just want to note how far off the mark genetic determinism actually is. To begin with, Huntington's disease is rare, and an isomorphic causal chain from genotype to phenotype is also rare. The official message of the scientific community is clear: the relationships between genes and visible traits or diseases are extremely complex. Environmental, developmental, and epigenetic forces can all play a role. The relation between genotype and phenotype is less a causal chain and more a web of intricate associations, in which precise causal chains and predictable outcomes are the exception rather than the rule. This complexity was underlined vividly by biologist William Gelbart (1991), who said that completing the HGP is less like finding the Rosetta Stone, and more like discovering the Phaistos disk—a set of as-yet undecipherable glyphs from a Minoan temple.

In the official rendition of genetic science, determinism is a vast oversimplification, operative in only a very few cases, and the tendency to generalize to all conditions and behaviors is a gross distortion. Still, simplistic stories of how genes shape us physically and behaviorally continue to occupy a central place in both scientific and popular culture. I will refer to this as the dogma of determinism. Examples are plentiful. There is a long list of conditions, diseases, and behaviors that are often portrayed as caused by genes. Not just cancers, heart disease, hypertension, alcoholism, hyperactivity, and Alzheimer's, but also stress, violent tendencies such as IED, risk-taking behaviors, aggressiveness, pyromania, shyness, and many others. There is even speculation about a gene for happiness, and my favorite so far—the "God gene." Dean Hamer (2004) argues that a self-transcendence mechanism is hardwired into us, for those of us with the VMAT2 polymorphism, like birdsong is hardwired into a goldfinch. Avoiding a total reductionism, Hamer finally eschews the definite article about halfway into his book and reverts to "a" God gene, conceding that there might be other genetic contributors to spirituality. But his focus on biological components and the language of inevitability remains strong throughout his work and terms like "hardwiring" are still prominent enough to be in the book's subtitle.

DETERMINISM AND THE HUMAN NEED FOR EXPLANATIONS

The appeal of determinism as a vehicle for self-understanding is not hard to identify. As a species we have a drive to know and understand who we are, where we came from and where we are going. Maybe humans have a gene for this drive to understand! (Or so a genetic determinist might claim). But whatever its source, we seem to need stories that capture our identity, that explain our origin and essence. Abraham Maslow (1943) even put it on his Hierarchy of Needs. And genetic determinism gives a clear response to these perennial questions of self-explanation.

This observation is not new to me. Dorothy Nelkin and Susan Lindee (1995) and others have written eloquently on the cultural appeal of genetic explanations. Nelkin (1991) points out that when E. O. Wilson's *Sociobiology* was published in 1975, *Business Week* ran an article entitled "The Genetic Defense of the Free Market." Capitalism is, presumably, a natural state of affairs because we are programmed for it. Barbara Rothman (1998, p. 13) has put it well and succinctly in saying that "Genetics isn't just a science. It's become more than that. It's a way of thinking, an ideology." I would add, a way of thinking in which the dogma of determinism is just under the surface, and sometimes clearly visible above it.

I want to push Rothman's notion of genetics as a way of thinking one step further and suggest that genetics has come to be a way of thinking with mythic proportions. I suggest that genetics functions like a myth, not of course in the sense of a fiction, the opposite of truth, but as something deeper than factual truth. Genetics is a myth in the sense of an allegorical or parabolic story that embodies an existential truth, an explanation of the meaning of things. As a student of religious behavior, I think human beings cannot lead their lives without some mythology, whether secular or explicitly religious. These mythologies can have many forms and implications. Some are insightful and beneficial, while others can be toxic to social life and community. One of our dominant cultural fables in this country, for example, is our self-portrait as rugged, self-reliant individuals who should be able to solve our problems with little or no social support. This is arguably the ideology behind current American health policy, which is brutally punitive to the unemployed and those in low-paying jobs. The myth of the Garden of Eden, to take another example, is a story of origins that satisfies at least two deep needs: a belief in exalted origins and a recognition of inevitable screw-ups, which explains why although humans are just one step down from divinity, we often end up in the ditch.

The work done for us by the mythic features of genetic determinism are apparent. They lead us away from the hard choices of social involvement and environmental regulation. If genes are the main actors, everything else is reduced to a supporting role at best. If learning deficiencies can be explained by genetic deficits of the brain, then the vexing problems of cultural deprivation, lack of educational opportunities, nutritional effects on attainment, etc., can be set aside. This may remind readers of the notorious best-selling book, *The Bell Curve* (Murray & Herrnstein 1994), which essentially argued that divergent results on standardized intelligence tests are best explained by genetic differences. In the end this means that no one is morally complicit in the outcomes.

I cite these examples not simply to emphasize that the dogma of determinism can play into some of our most perverted fears about differences between individuals and groups, but to point to the debilitating effect on ethics when choices are removed, and responsibility is allowed to go on holiday. Historian Mitchell Ash (1994) has written that the chief challenge that we face in the genetic era is not so much the abuse of genetic knowledge, but abuse from the discourse in which that knowledge is couched. Maybe the big hazard is not that we will be biologically manipulated, but that we will be manipulated by a mythic view of genetics and ourselves that is overly deterministic. The dogma of genetic determinism, combined with the dominant American values—militant individualism, a consumer-oriented, market-driven culture, and technological infatuation—seems like a recipe for truncating both personal and social awareness of responsibilities.

Perhaps a more complex and accurate understanding of genetics will be forthcoming in the future. Yet the sound-bite, sloganeering features of deterministic understandings of genetics will be difficult to displace. Even when scientists back away from the uni-directional, simplistic causal language of genes as "blueprints," they often refer to genes as "fundamental units" of life containing "all the information needed," or "the programs" for development. This language is not only psychologically comforting, as I suggest, but *familiar*—historically embedded in our scientific and cultural vocabulary. And here I draw on the work of Evelyn Fox Keller (1995, p. 93), who notes that we have inherited understandings of the gene that portray it as a combination of a "physicist's atom" and a "Platonic soul"—a mixture of building block and animating force. Telling stories that recognize complexity, with myths of indeterminacy and an openness in which moral agency plays a prominent role, will be a continuing challenge.

GENETICS AND THE MULTIDIMENSIONAL MYTHOLOGIES OF RELIGIOUS FATE

I suggested earlier that religious languages of fate and destiny were less damaging idioms of determinism, less threatening to ethics, than those of popular science or medicine. I now want to make good on that claim. I have noted in my own interviews with patients who have strong religious or spiritual orientations that they frequently speak from what seem like contradictory premises. For example, couples trying to decide on a controversial prenatal surgery for spina bifida sometimes say things like: "Well, it's all in God's hands," and at the same time believe they have a responsibility to make the best choice possible, and manifest this by learning everything they can about the benefits and hazards of the procedure (Rothschild, Estroff & Churchill 2005). On the surface this looks like an impossible combination of divine omnipotence and human moral agency. I have also noticed that petitionary or intercessory prayer often follows a similar illogic. People pray for things, like being cured of an illness, while at the same time fervently believing that the shape of their lives and the timing of their deaths have been already decided, and it is just a matter of playing it out. If we are to take this seriously, and not as mistakes of logic, we will need to accommodate a kind of moral agency that sees responsible choice as legitimate even when it appears to be superfluous, or impossible. Here the felt sense of responsibility to do the right thing, even if only to pray the right kind of prayer—even a prayer seemingly made redundant by a metaphysic of fate— is not just an unthinking residual habit, but seen as a genuinely obligatory action. Here the felt experience of ethics goes beyond not only what Judith Thomson would accredit as the proper range of responsibility, but even beyond what someone like Simon Blackburn would argue are the theoretical requirements for ethics to be meaningful. And my experiences have also been observed by other researchers.

In her studies of patient perceptions of genetic diagnoses, Tina Harris and her colleagues (Harris et al. 2009) interviewed 90 people about the uses of their religious frameworks as a lens for understanding genetic illnesses. They found, not surprisingly, that there were a wide range of views. For some their religious determinism was just as absolute and uni-dimensional as the casual, off-hand determinism espoused by some scientists and clinicians. Some patients said: "If it's mean to be, it's meant to be," meaning that neither health behaviors nor religious choices make any difference. "The gene means the disease," or, as Harris and colleagues summarized this position, "the presence of the gene means the person *already has* the disease" (2009, pp. 25ff). Yet these same people reported that they continue to be religiously observant. A similar finding is reported by Mary White (2009), citing the work of Anita Kinney and colleagues, who found that among women identified as being at risk for BRCA1, those who thought of God as the "locus of control" were no less likely to pursue testing than those who did not.

For other people in Harris's study, either health choices or religious actions, or both, were thought to reduce the risk of a disease developing. They believed that being proactive can make a preventive difference, which implies a significant range of responsibility. And for still others, there was thought to be no link between health or religious behaviors and genetic illness. This last group seemed to be saying that prayer can change you, but not your cancer. This theme is as old as the Stoics and as fresh as the modern Existentialists, both of whom insisted that sometimes the only thing human beings have a choice about is their attitude to life's misfortunes, but that this choice of attitude, far from being insignificant, is a defining feature of ethics.

Harris ends the report of her study in a practical way, suggesting that patients often hold mutually exclusive understandings—without calling them misunderstandings—which can pose challenges for the health professionals trying to assist them. The answer, she argues, is the development of culturally sensitive models for patient care.

Culturally sensitive patient care is surely important, but my interest lies in a different direction. I am concerned with how purposive, intentional actions can cogently fit with expressions of determinism. If people hold mutually exclusive understandings, what does it tell us about the nature of ethics, not as an academic field, but as a human endeavor? At a minimum it means that responsibility can still find traction even when the larger metaphysic is expressed as fate or destiny. I have argued that one way of accounting for this is to expand the notion of truth in ethics to include what I have called mythological truth. Mythology responds to a felt depth about ourselves and others that no scientific response can satisfy. Rothman (1998) has analyzed the ways that genetic thinking provides a powerful way of imaging our futures. Interviews with patients show that genetics is one way of imagining the future among other ways, and that deterministic ideologies in either science or religion often do not preclude a future in which responsible choices play a role.

A true-believing and thorough-going reductionist could still argue that it is theoretically possible that genetic determinism could, in the end, be a complete account of human life. In this view it is only a matter of time until science will eventually trace all the intricate causal pathways in the proteomic-environmental chain, and we will be, as it were, fully described and completely predictable. But even if this were true, which I doubt, humans would still have to make sense of their status as fully described and completely predictable beings, that is, attribute meaning to it. To be sure it could be argued that—as I jokingly suggested earlier—this drive to make sense of ourselves is itself no more than the expression of a gene or set of genes working in concert. But then the explanation begins to strain credibility. Is my interest in writing papers with discursive footnotes also the product of a genetic pattern, rather than a personal aesthetic and scholarly preference? But maybe my aesthetic and scholarly tendencies are likewise genetic products. It is easy to see where this line of argument goes, and why it is unsatisfactory. At some point the explanations become hopelessly contrived, and some degree of genuine choice has to enter into the scene. The real picture seems far more complex than infinite layers of determinisms.

Neuroscientist Michael Gazzaniga (2005, pp. 100 ff), discussing the problem of accounting for free will, says that what is needed is a sufficiently stratified understanding of human beings. "Brain" and "neuron" are biological concepts, whereas "responsibility" is a social concept, "something that exists only in the context of human interaction." This means that no brain scan or *f*MRI will be able to capture free will, or depict moral agency. To look at a brain scan for a picture of responsibility or free will is like examining the scan of a baseball hoping to find a picture of a knuckleball or a slider. A parallel argument can be made for the effort to reduce freedom of choice to the actions of genes. Genes, at best, can only show the biological preconditions for the emergence of human moral characteristics, just as adequate blood flow to my prefrontal context is a necessary, but not a sufficient condition for me to write this paper. The problem is what Michael Polanyi (1969) would call a transfer of concepts and frameworks from one, lower level of ontology to a higher and more complex level, where they lose explanatory power. Brains are not minds; genes are not persons; predispositions are not choices. It is mistaking a part for a whole. Science can explain neurophysiological mechanisms; it cannot explain realms of human meaning to which both responsible choices and spiritual experiences belong.

If human beings could be said to be programmed or hardwired for anything it is for a remarkable plasticity—biologically, psychologically, ethically, and spiritually. If we transfer the model of a stratified ontology to the statements of patients as they contemplate and deal with genetic diagnoses, we might say that when they discuss fate or destiny, they may well be talking intelligibly about a cosmic sense of reality that exceeds their understanding, while at the time, they feel comfortable in a personal and immediate awareness of moral agency. No contradiction. Just an appreciation of levels of being and layers of meaning which both scientific and religious dogmas often fail to appreciate.

CONCLUSION

In bioethics and in the health professions more generally we talk a lot about making responsible choices in the face of good information, about learning the facts to help ourselves and others be realistic and well-informed in our decisions, all in the service of a robust autonomy. This is clearly good as far as it goes, but it does not go far enough. In the practical world it seems that determinism does not preclude responsibility, and that moral agency can be an effective force even in the face of the impossibility for change. So in addition to educating people for autonomous choices we in health care also need to be listening for how these choices reflect various layers of responsibilities that move beyond traditional logic and the scientific framework. A notion of ethics that tilts too heavily toward volitional and explicit cognitive processes cannot speak to those dimensions of ethics that are less about choosing and more about sense-making, about both discovering and being responsive to a meaning when things cannot be changed, about deciding on which level of life we will live. Despite the important focus of contemporary bioethics on explicit decisions, the ethical choices of patients are often in this more tacit realm.

Religious determinism, often treated as unscientific or worse, is for many people not a barrier to ethics and responsible choice at all. One could even argue that religious notions of fate or destiny are in fact a spur to a heightened sense of responsible choosing, insofar as people in dire straits often pray to be adequate to the inevitabilities they face.

This observation is not far from something that Max Weber (1958) noticed about Protestant, Reformed traditions of Christianity. People who believed in foreordained election to paradise or damnation seemed to work very hard to demonstrate that they were in fact among the elect, Weber argued, with material possessions serving as a sort of outward and visible sign of an inward and invisible grace. Hard work and its material rewards were embraced as signs, even though it was also believed that grace was foreordained and could not be earned. Weber's thesis is a classic indication that religious/spiritual affirmations of fate often have a robust place for the ethics of responsible choice.

The phenomenon I describe here and associate with religious/spiritual interpretations may be more broadly characteristic of human frames of meaning. For example, sociologist Michele Easter (2009) says that her work to date indicates that people with eating disorders feel that genetic ideas reduce the stigma, increase the sense of fatalism, but can also be "empowering" in terms of personal agency. So even outside of religious/spiritual frames of reference, persons may embrace complex and layered understandings that defy linear logic and the neatness of professional theorizing, whether that theorizing is scientific, theological, or bioethical.

In summary my argument is this: while many theories of responsibility emphasize voluntariness, the more burdensome sorts of responsibility may be those we did not choose; that to lessen that burden, deterministic portraits of ourselves are often quite appealing; that the prevalent popular, historic, and clinical assumptions about genetics often verge on determinism; that religious/spiritual interpretations of determinism, expressed as "fate" or "destiny," are less damaging to ethics, since whatever our theories, fate and choice seem to be complements in many religious/spiritual sensibilities, rather than alternatives.

Perhaps this is another example of what theorists in ethics can learn from close observation of human practices.

ACKNOWLEDGMENTS

I gratefully acknowledge the assistance of my colleagues at the Center for Biomedical Ethics and Society at Vanderbilt University for their many helpful suggestions.

REFERENCES

American Psychiatric Association (2000). DSM-IV. Washington, D.C.: APA. pp. 663–667.

Ash, M. (1994). Human capital and the discourse of control. In R. Weir, S. Lawrence & E. Fales (Eds.), *Genes and human self-knowledge: Historical and philosophical reflections on modern genetics*. Iowa City, IA: University of Iowa Press.

Blackburn, S. (2001). *Being good*. Oxford: Oxford University Press.

Brock, D. (1994). The human genome project and human identity. In R. Weir, S. Lawrence & E. Fales (Eds.), *Genes and human self-knowledge: Historical and philosophical reflections on modern genetics*. Iowa City, IA: University of Iowa Press. 18–33.

Churchill, L. R. (2009). Religion, spirituality and genetics: Mapping the terrain for research purposes. In J. B. Fanning & E. W. Clayton (Eds.) *American Journal of Genetics* 151C (1): 6–12.

Easter, M. (April 3, 2009). Drawing a line between choice and disease: Early findings from dissertation research. Presented at the Cultural and Political Sociology Workshop, University of North Carolina at Chapel Hill.

Gazzaniga, M. (2005). *The ethical brain*. New York: Dana Press.

Gelbart, W. (1991). A vision of the grail. In D. Kevles & L. Hood (Eds.), *The code of codes*. Cambridge, MA: Harvard University Press. 83–97.

Hamer, D. (2004). *The god gene: How faith is hardwired into our genes*. New York: Doubleday.

Harris, T. M., Keeley, B. Barrientos, M., Gronnvoll, J., Landau, C., Groscurth, L., Chen, L., Cheng, Y. & Cisneros, J. (2009). A religious framework as a lens for understanding the intersection of genetics, health and disease. In J. B. Fanning & E. W. Clayton (Eds.), *American Journal of Medical Genetics* 151C (1): 25.

Hubbard, R. & Wald, E. (1993). *Exploding the gene myth: How genetic information is produced and manipulated by scientists, physicians, employers, insurance companies, educators, and law enforcers*. Boston: Beacon Press.

Kaplan, J. (2000). *The limits and lies of human genetic research*. New York: Routledge.

Keller, E. F. (1995). *Refiguring life: Metaphors of 20ᵗʰ century biology*. New York: Columbia University Press.

Kevles, D. & Lappe, M. (1995). Eugenics. In W. T. Reich (Ed.) *Encyclopedia of bioethics* (Rev. Ed.), vol. 2. New York: Macmillan.

Maslow, A. (1943). A theory of human motivation. *Psychological Review* 50(4): 370–396.

McGee, G. (1997). *The perfect baby: A pragmatic approach to genetics*. Lanham, MD: Rowman & Littlefield, pp. 60–65.

Murray, C. & Herrnstein, R. J. (1994). *The bell curve*. New York: The Free Press.

Nash, J. M. & Thompson, D. (1989, March 20). The gene hunt. *Time* 133:12.

Nelkin, D. (1991). The social power of genetic information. In D. Kevles & L. Hood (Eds.), *The code of codes*. Cambridge, MA: Harvard University Press. 177–190

Nelkin, D. & Lindee, S. (1995). *The DNA mystic: The gene as a cultural icon*. New York: H. H. Freeman.

Polanyi, M. (1969). *Knowing and being*. Chicago: University of Chicago Press. 225–239.

Rothman, B. K. (1998). *Genetic maps and human imaginations: The limits of science in understanding who we are*. New York: W. W. Norton. 173.

Rothschild, B. Estroff, S. & Churchill, L. (2005). Experimental maternal-fetal surgery: The cultural calculus of consent. *Clinical Obstetrics and Gynecology* 48(3): 574.

Silber, J. (1967). Being and doing: A study of status responsibility and voluntary responsibility. In J. M. Edie (Ed.), *Phenomenology in America*. Chicago: Quadrangle Books.

Thompson, L. (1994). Communicating genetics: Journalists' role in helping the public understand genetics. In In R. Weir, S. Lawrence & E. Fales (Eds.), *Genes and human self-knowledge: Historical and philosophical reflections on modern genetics*. Iowa City, IA: University of Iowa Press. 104–121.

Thomson, J. J. (1971). A defense of abortion. *Philosophy and Public Affairs* 1(1): 47–66.

Weber, M. (1958). *The Protestant ethic and the spirit of capitalism*. New York: Charles Scribner's Sons. 104 ff. (First published 1905.)

White, M. (2009). Making sense of genetic uncertainty: The role of religion and spirituality. *American Journal of Medical Genetics* 151C (1): 72.

7 Responsibility versus "Blame" in Health Communication

Where to Draw the Lines in Romancing the Gene

Roxanne Parrott

Individual choices about health are not autonomous but are influenced by powerful political, social, and market conditions. Genomic health care linked to the mapping of the human genome is no exception. Support for the Human Genome Project relating to goals associated with understanding all of the genetic material contained in an organism or a cell was debated in scientific, political, and personal realms. Reports about these conversations appeared in newspapers and scientific journals. Sometimes, the debates themselves were aired on television and had broad public audiences. Scientific understanding of the genome was presented as a path toward disease prevention represented as opportunities to implement findings in translational research intended to improve lives (Friedrich 2004). These opportunities include new drugs to treat common diseases and dosing guidelines based on individual genetic make-up, pharmacogenomics, and food and vitamin supplements aimed at nutritional status based on individual genetic make-up, nutrigenomics.

Health communicators, as both applied communication and translational researchers who derive understanding about the effects of communication processes in public health and medical interaction realms, are challenged to improve health and risk communication (Parrott 2008). The primary goal is often to support informed decisions. In the case of genomics, a core aim relates to informed decisions about the role of multiple genes for common conditions such as cancer, diabetes, and heart disease. Interdisciplinary applied communication and translational research endeavors have significant opportunities to influence guidelines about the responsible conduct of research in this era of genomics, public health, and health care. These "rules," sometimes formally adopted as health policy and other times informally scripted within societal groups, make individuals and society accountable for acting or failing to act on behalf of personal and public well-being. Accountability is the allocation of responsibility and/or liability for health and health care to individuals, professionals, and institutions. As a result, the principles linking genomics, public health, and medical care also set up a dynamic linked to *blame* in which individuals and society become accountable for harms to themselves and others.

We, as individuals, are held accountable for health harms ranging from those inflicted on ourselves through actions such as smoking, and driving under the influence of alcohol or drugs. Automobile insurance companies may have a policy that charges different premiums depending upon whether the policyholder is a smoker or not. These companies may also have rules about what coverage they pay in the event that an accident occurs and the insured is not wearing a seatbelt, or in the case of a motorcycle accident and failure to be wearing a helmet. These rules are the formal statements of accountability associated with responsibility and implicitly reflect blame linked to harms. Some fear that such accountability issues will cross over into the genetic realm. While one has freedom to choose to wear a helmet, one has no such volition in choosing one's genes. What happens if information about a particular gene predisposing someone to tobacco or alcohol addiction is used to support charging higher rates for car insurance, seeming to *blame* someone for his or her genes even in the absence of behavioral evidence that she or he is smoking or excessively consuming alcohol?

In the U.S., the passage of the Genetic Information Nondiscrimination Act (GINA) of 2008, 13 years in the making, explicitly states that limitations aligned with health insurance do not apply to life insurance, disability insurance, or even long-term care insurance (http://www.genome.gov/Pages/PolicyEthics/GeneticDiscrimination/GINAInfoDoc.pdf). While car insurance is not referenced in the document, it does not take a great deal of imagination to identify strategies to leverage use of genetic information for insurers' profit in auto and the realms of life, disability, and long-term care insurance. Efforts to apply the use of genetic tests to avoid disability insurance payments have been identified in the past. One case, for example, involved the Burlington Northern Santa Fe Railroad (BNSFR), which relied on the company physician's advice, who relied on the recommendation of a diagnostic company's representative, to obtain blood for DNA testing from employees seeking disability compensation for carpal tunnel syndrome that occurred on the job (Clayton 2003). The federal Equal Employment Opportunity Commission stopped the practice in this case, but questions remained about why testing was pursued in light of the cost associated with testing and the rarity of the genetic disorder they were testing for, an occurrence of three to 10 persons in every 100,000 (Clayton 2003).

The BNSFR case illustrates a concern that genetics tests may be adopted without having demonstrated their benefits, leading to severe consequences for individuals, families, and communities (Burgess 2001). Even for individuals, as compared to employers, GINA does not prevent health insurers from using disease *manifestation* linked to genetics as a basis to determine eligibility or premium rates. To consider this reality, consider the case of venous thromboembolism (VTE), blood clots, which affect hundreds of thousands of us at all ages. There are an estimated two million cases of VTE in the U.S. annually with a mortality of 60,000 (Hirsh et al. 2001). Medical science has shown that several genes, with others likely to be found in the future, increase the possibility of developing a blood clot (Crowther & Kelton 2003). What happens if information that hundreds of those who have experienced blood clots have one or more of these genetic contributors to blood clots is used to justify refusal to hire them in jobs requiring long periods of sitting, such as being a trial lawyer or an air traffic controller, or to charge higher health and life insurance premiums? Is such *blame* warranted in the name of the public's good—costs linked to genetic make-up?

When awareness informs our own choice in these matters, such that we choose not to have a career that requires us to sit for hours without leaving our post, it defines informed decision-making. Science leading to policies which limit our options defines discrimination. Definitions of responsibility versus blame in efforts to participate in applied communication and translational research relating to genomics is a core ethical dilemma for health communicators. Two broad issues are relevant in discussions relating to strategic efforts to communicate about genomics, public health, and health care. What requirements should be met relating to the readiness of medical research to be translated into potential interventions before proceeding? What barriers to health communication translational efforts limit its effectiveness, leading to unintended effects?

REQUIREMENTS TO UNDERTAKE TRANSLATIONAL RESEARCH

Behind decisions to pursue translational research are decisions about the readiness of evidence to form interventions for the betterment of health and, admittedly, the likelihood of satisfying profit motives to justify the initiatives. Toward that end, not all medical research is judged to be equal. Genetic testing decisions, for example, have been categorized based on both their clinical validity and the efficacy of available treatment (Burke, Pinsky & Press 2001). The broad goal in translational research is to enable individuals and society to make responsible decisions when allocating resources related to health and health care. This requires identifying the duties associated with both our personal and society's social obligations related to genomic public health and health care. How this will play out in relation to genomics in clinical and public health settings requires some projection and also benefits from historical reflection.

CONVERGENCE RELATING TO THE EVIDENCE

Long before the era of genomic health care, communication about one gene predicting the likelihood of inheriting one disease was revealed to be wrought with complexities. A single genetic test result gives us, our family, and the medical care team *one* piece of information about our health. When testing for a gene relating to cystic fibrosis or sickle cell disease, for example, the genetic test result provides enough information to guide communication about the likelihood that someone will inherit or carry the condition. Counseling relating to situations where someone is a carrier of a gene relating to these single gene conditions affects many decisions relating to important life issues. Who will the carrier tell about the condition, an invisible one, and when? How will it affect decisions about having children? Multiple genetic contributors to a health condition multiply the complexity relating to communication about genes and health.

Before pursuing applied health communication and translational research linked to genomics, there must be agreement regarding the medical research related to genetic contributors linked to a chronic condition. Medical research has converged in relation to evidence, for example, that genetic variations contribute to hypercholesterolemia (Marteau, Senior, Humphrises et al. 2004), the motivation to use drugs (Goldman, Oroszi & Ducci 2005), to smoke (Lerman et al. 2002), variance in how pain is experienced, and blood clotting risk (Crowther & Kelton 2003), to name but a few of the published findings. There must also be convergence relating to the possible benefits to be derived by translating the research into practice. Too often, miscommunication and misunderstanding often arise as a result of misattribution associated with responsibility in health communication. Genomic health care compounds this likelihood.

As an example, consider the rising incidence of obesity in childhood. Discourse about responsibility includes discussion about: 1) the child's overeating and not exercising, 2) the parents' role in making such foods accessible, 3) food manufacturers' role in producing high fat foods, 4) public education's role in reducing the number of physical education classes, 5) pediatricians' role in discussing long term negative health consequences, and 6) genes' role in making some more susceptible to obesity and even to overeating in the first place. Each of these, and in turn, any combination of them, guides the amount of energy and other resources an individual, professional, or institution will devote to childhood obesity prevention. To translate the research into efforts to improve well-being, reducing the incidence of obesity, health communicators might focus on the child, the parent, food manufacturers, public education, *and* pediatricians as targets for communication about the issue. In doing so, a more valid representation of the situation is acknowledged than to focus on only one among the group. In doing so, messages could focus on food and nutrition, exercise, and opportunities for both, and all messages could encroach into discussions about the role of genes.

A primary argument emerging in support of genomic health care is that individuals can make more informed decisions about care and be motivated to behave in health promoting ways. This likely depends on how genetic information is communicated and communicating that medical innovation has risk. Research has shown that participants who tested positive for a genetic mutation relating to hypercholesterolemia believed *less* in the effectiveness of diet, their personal response to high cholesterol levels, to reduce cholesterol levels (Marteau, Senior, Humphrises et al. 2004). The latter group may be motivated to follow a low cholesterol diet due to the positive genetic test result but also believe that an additional response, such as cholesterol lowering medication, will be needed to promote their well-being. Or, those who test positive may believe that no matter what diet they consume, their levels of bad cholesterol will only be controlled through cholesterol lowering medication.

The medical research and behavioral science evidence suggests just telling someone their genetic status without further conversation about ways to apply the information to their lives leaves opportunities for harm which may lead to blame. One survey of 186 smokers in London, for example, found that 65% believed that they would be motivated to quit smoking if they were found to have genes related to diseases involving multiple genetic contributors and smoking (Sanderson & Wardle 2005). That leaves 35% without the belief that the information would be motivating. The lay public may be motivated and able to sort through contested meanings of genes, but there is likely to be a wide range of models used to represent their understanding. In my own research (Parrott et al. 2004), our team discovered that some of us are simply *Uncertain* about the roles of personal behavior, social support, and religious faith on genes in health, while others believe that these factors all contribute equally to how genes express themselves in individuals' health, a more *Integrated* epistemological framework. A third group doubts the role of religious faith for genetic expression in health, and is also uncertain about a role for social support, but perceive personal behavior to have an important role, suggesting that this group places primary emphasis on *Personal Control* for the role of genes in health. A fourth group believes that the role of genes for health has little relationship to personal behavior, religious faith, or social support, indicating a likely emphasis on the genes one is born with in deriving their belief system about genes and health, characterizing the more classic *Genetic Determinists*. These models suggest that health communication translational researchers will propose different approaches to communicating about genomics and health, some of which may conflict with satisfying a profit motive to warrant its pursuit.

SATISFACTION OF A PROFIT MOTIVE

With the ability to produce proteins and enzymes, gene therapy becomes a topic for discussions inside and outside of research and clinical settings. Gene therapy is the use of protein and enzyme products to replace, manipulate, or supplement genes that relate to or are less resistant to disease. Gene therapy options are being aggressively pursued, for example, in research for preventing cystic fibrosis and sickle cell disease. The research has been successful in studies for sickle cell using mice, but even here is far from prime time for use as human therapy (Seppa 2001). Methods to automate DNA sequencing have been designed. These include bio-imaging systems such as GeneGnome (http://www.syngene.com), which automate the process of imaging chemiluminescent samples, providing accurate results that are easier to use. So the ability to accurately *tell* a woman that she *has* BRCA1 or BRCA2 mutations, for example, has been enhanced. The options relating to this information, however, do not include replacing, manipulating, or supplementing the genes—gene therapy. So, once more, communication about these matters is a slippery slope.

The reality associated with the promise of profit and benefits for genomics and human health will often depend upon the lay public's decisions about participating in clinical trials linked to genomics now with accrual of human health benefits far in the future. For health communication translational researchers, the goal thus focuses on recruitment to clinical trials and medical research when there may be no personal reward for those who agree to participate. The public will be solicited to provide our cooperation in giving lifestyle information, family health histories, personal medical information, and biological specimens (Friedrich 2004). Why? So that genetic databanks might be assembled with linkages to the multiple determinants of health. Because such data exists for some groups, some groups have been identified as being at increased risk for having genetic mutations being related to some disease.

A centralized and accessible national repository of biospecimens focused on supporting genetic and proteomic research has been proposed. The national biospecimen network (NBN) blueprint (Friede, Grossman, Hunt, Li & Stern 2003) commissioned by the National Dialogue on Cancer and the National Cancer Institute, for example, argues for this level of organization relating to research associated with genes and health. Patient advocates and consumers, however, have objected to the NBN's emphasis on researchers' needs (p. 7 of the summary of public comments in response to the blueprint). The blueprint outlines barriers to taking an approach that would involve the donors more directly, which are ones common in many clinical trials. These include the reality that the deidentification of tissues makes it difficult to recontact donors. Also, the progression of a patient's disease makes treatment irrelevant based on new information and the reality that validated results would be unlikely to be available during the course of a patient's active disease (pp. 17–18 in blueprint). Once more, health communicators find the reality of representing this situation to be a challenge in the wake of trying to recruit participants to any type of clinical trial, but magnified in the wake of unknowns linked to genomics and subsequent health policies.

In the process of cooperating by providing personal information, we risk personal, social, employment, insurance, and criminal discrimination (Parrott et al. 2005). One five-year study, the Genetic Discrimination Project conducted in Australia, found that about 100 (10%) of 1,000 people had experienced some form of discrimination, with 14 people being refused insurance. Who's to blame? There is a rising tide of public resentment because medical technology and therapies seem to benefit others but not us. The truth, however, is often a very different one. The truth is that medical technology and therapies often are a reality for *no one*. They simply have not been discovered, invented, designed, approved, or produced yet. In the case of pharmacogenomics, where arguably genomics and health are converging to satisfy a profit motive to prescribe based on individual genetics, several steps lead to the development of genetic tests for the genes linked to such chronic conditions as diabetes or cancer, and blood clotting tendencies—with profitability ranking high on the criteria.

Warfarin therapy for treatment of blood clotting disorders, for example, has been found to be associated with several gene variants which reduce the dose of the drug needed in patients, explaining 50–60% of the variance in warfarin dose requirements for Caucasians and Asians (Cavallari & Limdi 2009). That means, of course, that 40–50% of patients do not benefit from this knowledge, but in view of the reality that 1.5 million people in just the U.S. take the medication every single day, there is significant costs associated with the testing and variation in drug response that can be reduced via identifying factors that contribute to this situation. Thus, the profit and health care savings to be generated warranted developing tests to identify the genes that medical evidence cites as contributors and administering to those whose international normalized ratio (INR) scores fluctuate widely.

Other genetic variants contribute to the efficacy of other prevention aims, including, for example, the efficacy of bupropion used for smoking cessation (Lerman, Shields, Wileyto, Audrain, Pinto, Hawk, Krishnan, Niaura & Epstein 2002). Cases thus exist where convergence relating to scientific discoveries and the profit motives to support linking genomics and health has been realized with sufficient validity and reliability to warrant the involvement of health communication translational researchers. In these situations, health communication translational aims will focus on avoiding conflicting messages and deriving consistent communication to contribute to informed decisions. Barriers to achieving these aims can also be considered during strategic planning phases.

BARRIERS TO TRANSLATIONAL RESEARCH

While physicians often are a resource for health information and medical decision-making, they may not be aware of all the options available. In this rapidly changing era of genomics, public health, and medical care,

physicians are challenged to keep abreast of new knowledge and developments linked to available testing that may have relevance for their patients
(Kelly, Love, Pearce, Porter, Barron & Andrykowski 2009). Moreover, 21st
century media paradigms often bring medical research information to us
long before we are scheduled to see our doctor. The utility of the information often depends on our ability to sort through contested meanings. In
one survey, 87% of 649 participants indicated a desire to be tested for
genes relating to cancer but also conveyed little understanding of the implications of test results (Andrykowski, Munn & Studts 1996). Randomized
controlled trials of strategies to counsel women about breast cancer risk
relating to BRCA1 and BRCA2 genes support the importance of providing patients with information to guide testing decisions (Wang, Gonzalez,
Milliron, Strecher & Merajver 2005). Science and media literacy in the
genomic health domain predict the likelihood that such information will
be beneficial in supporting informed decisions.

SCIENTIFIC ILLITERACY

As we move into the new era of genetic medicine and pharmacogenomics, the need to grapple with health literacy will become even more
compelling. If concepts such as "teaspoonful" or "angina" are baffling
to a significant majority of the population in the US and around the
world, how are we going to convey the complex genetic information
that represents both the promise and the potential peril of the dawning
medical genetics age? (Giorgianni 1998, p. 1)

Basic understanding of human genetics appears paramount to informed
choice, informed consent, and shared decision-making in this era of ever-
widening genomics in public health and clinical practice. Yet, not surprisingly in view of the challenges posed to medical experts to keep up with
the research, lay audiences struggle to understand human genetics. Some
researchers conclude that patients *resist* risk information, and risk communication may lead patients to make decisions that they would *not* make
with more accurate understanding (Gurmankin, Domchek, Stopfer, Fels
& Armstrong 2005). The reality seems to be more likely related to knowledge and the ability to connect understanding to informed decisions. One
survey of 1,216 Finnish adults regarding their knowledge of genes and the
relationship of genes to disease, for example, found 15% answered less
than seven of 16 questions correctly, while 62% knew the right response
to between seven and 12 of the items; more knowledge was directly related
to greater reservations about the wholesale adoption of genetic technology,
testing, and screening (Jallinoja & Aro 2000).

Human genetics information is complex, abstract, and technical. One
study found that when parents received information about a prenatal

diagnosis of sex chromosome abnormality, materials just copied from a medical textbook, 26.2% of the parents reported that the reading materials were not helpful because they were too scientific and clinical (Petrucelli, Walker & Schorry 1998). Moreover, efforts to communicate health information to the public frequently depend upon the use of numbers rather than verbal expressions. Numeracy is the ability to use basic probability and numerical concepts, with research findings that individuals are able to more accurately calculate their benefit from mammography as numeracy increased (Schwartz, Woloshin, Black & Welch 1997). In a genetic counseling setting, one study found that when numeric risk information was presented with the denominator as 1,000 rather than 100, pregnant women were much less likely to assign the correct percentage equivalent (Chase, Faden, Holtzman, Chwalow, Leonard, Lopes & Quaid 1986). Women presented with rates versus proportions in assigning risk for Down syndrome were better able to understand rates (Grimes & Snively 1999).

The language used in discussions of science and health challenges the public across multiple domains but is proving to be specifically problematic in efforts to move genomics to public health communication and medical interaction domains. Humans all share mostly—99.9%—the same genes, but we may have different versions, or alleles, of the same gene. The version of a gene that is less common is referred to as a mutation. In one study I have been involved with, 243 lay participants were asked to rate their perceptions of the terms mutation, alteration, variation, and change (Condit et al. 2004). Mutation was judged to be a more negative term when compared to all other terms. Mutation was viewed as *bad, unhealthy, not normal*, and *undesirable*. An alteration, on the other hand, was perceived as "intended" rather than "unintended" when compared to other terms. This may foretell the types of expectations forming around genomic health care. Too often, we may assume that tests are available for the many genes being linked to many diseases, believe that we can "alter" these genes linked to disease, and desire to do so because having a genetic mutation is met with feelings that somehow we are not as good as someone who has not been told they have a specific mutation. This is a barrier to undertaking translational research in relation to genomics. It suggests that information ought to be included about the fact that *healthy* people have genetic mutations. Health communicators may need to communicate that these differences may actually be positive in some cases and may promote human adaptation and survival.

Another barrier relating to science literacy and health communication translational activities relates to inconsistent discussion of the role of genetic contributors. When looked at one by one, this misses the meaning of genomic health care in which one gene, one disease, is *not* the way to think about the role of genes for health. Having a single mutation does not mean you *will* develop a disease. Not having the mutation does not mean you *will not* develop a disease. Lack of understanding about disease causation and the human genome is an area where the public's literacy levels suggest the need for concrete and specific content to be included in efforts to communicate. For lay audiences, reactions to several phrases that might be used to convey content about the perceived level of risk related to a genetic contributor to chronic disease revealed differences to guide health communication translational efforts, beginning with the foundation of understanding that needs to be communicated. The phrase "has a gene that causes" produced the highest levels of perceived risk in research comparing it with "a gene for," which held higher risk perceptions than "a family history of" did. Further analyses revealed that the two expressions containing the word "gene" did not produce statistically significant differences in perceived risk, but both were associated with greater risk as compared to "family history" (Condit & Parrott 2004).

The latter brings yet another realm of concern for health communication translational efforts related to genomics. There quite simply exists too little awareness of family health history and its role for medical decisions, including diagnostic and treatment. This has been persistently demonstrated over the past several years, with some of the research I have been involved in revealing that among 717 adults, 70% rarely or never discussed family medical histories in their families (Weiner, Silk & Parrott 2005). Health information brings health and heritage front and center. Family members have been called "the medium through which inheritance flows" (Finkler 2000, p. 3). In this era of naming genes in health, we may have concerns that genetic differences will be used to disadvantage some families as compared to other families who do not have the genes selected for scrutiny (Juengst 1999). As a result, there may be strict boundaries around disclosure of health information in families, as family members may work in concert to "conceal" their histories from outside members. Risk communication related to naming genes in health requires promoting awareness of our family health history.

Another barrier to use of family history to promote prevention relating to genes and health is the absence of such information in medical charts. This gap may be due to our own lack of awareness but may also be due to the failure of doctors to seek the information or our unwillingness to disclose for reasons previously discussed in relation to our failure to talk about family health history in our families. It is not enough for us to begin to identify relevant details of our family health history in our families. We must also be willing to disclose these details to our medical caregivers, and they must accurately record them in our charts. In one study of 995 new patient charts, involving 28 different primary care physicians, cancer information about family history was collected for 679 or 68% (Murff, Byrne & Syngal 2004). So, approximately 70% of the patients were able to give the information that they had a family history of cancer and it was recorded in their charts. In 414 of the 679 cases—61%, specific information about the relationship of the family member to the patient whose chart was reviewed was also recorded (Murff et al. 2004). Whether it was a parent, a grandparent, a sibling, or some other relative was included in the patient's chart. It is not clear in Murff et al.'s (2004) study of patient charts, however, whether the doctor failed to report the relationship of the family member in the other 265 cases, or whether a patient did not know *who* exactly had cancer in the family, only that someone did. These details will be increasingly important as genomic health care decisions depend upon a doctor's ability to recommend appropriate genetic tests, which depends upon our ability to tell the doctor our family health history. Promoting awareness becomes a core aim for health communication translational researchers.

In the Murff et al. (2004) study, when first degree relatives were affected, including a parent or sibling, the age of the relative at diagnosis with cancer was more often included as part of the record than in those cases where it did not involve a first degree relative. Once more, this may be because we are more likely to remember when a parent or sibling has such a diagnosis and have the ability to identity the age of a parent or sibling when they were diagnosed with cancer. Again, we cannot be certain that this is the situation, as doctors may be more likely to record information about age for first degree relatives because they regard the relationship to be more predictive of our health status. So both participants in the medical history-taking component must do their part in this era of genomic health care. We must know our family health history and be willing to disclose it in medical settings. Doctors must ask for and record that history, or record the fact that a patient did not know and the doctor indicated intention to follow-up on the next visit to obtain such information.

MEDIA ILLITERACY

> The bad news is that there appears to be a disconnect between the value scientists and doctors place on the information disseminated by mass media and the value placed on it by viewers and readers. Doctors routinely label a significant portion of health and medical coverage sensationalized, incomplete, inaccurate, unbalanced, or misleading . . . The worse news is that individuals come to mass media information and risk messages with an impoverished level of science literacy. (Rodgers 1999, p. 21)

Many of us learn about scientific discoveries through the media and share the information with others or act upon the information in hopes of benefiting our health. Media literacy is having the skills to critically analyze media messages. This includes the ability to understand techniques, technologies, and institutions which produce mass media. These include news organizations, entertainment industries, public relations firms, and advertising agencies. Two persistent problematic patterns appear in media translations of the science of health. One is content inaccuracies and the second is a lack of detail in the message, a bias of omission. One study of 116 popular press reports about mammography and breast cancer, for example, found that 113 cited a scientific study, but only 60 such studies could be found due to a lack of sufficient detail to locate the research (Moyer et al. 1995). Beyond such omissions, the researchers also found 42 content inaccuracies—testimony to the need for improved health science and numerical literacy among science reporters, as well as the general public. The addition of a phrase such as, "According to a study published in the *Journal of the American Medical Association* this week, Harvard medical scientists, Jane Smith and Joe Brown, found . . ." would enhance the ability of consumers and others to locate the research. The researchers recommended that reports include enough information that the scientific studies may be located, news agencies use fact-checking, and medical researchers train scientists to prepare for interviews about their work (Moyer, Greener, Beauvais & Salovey 1995). While not limited to media translations, these problems are most often identified in media.

Media theory and research also suggest a wide variety of stereotyping effects arising from the use of atypical exemplars for illustrating specific health issues. A growing body of research on exemplification effects indicates that atypical exemplification in media reports leads to skewed perceptions of social reality (Zillmann & Brosius 2000). Perhaps the first broad media coverage about genes during the era of the Human Genome Project was stories about breast cancer and the BRCA1 and BRCA2 mutations (Henderson & Kitzinger 1999). News stories about a "breast cancer gene" contributed to women's beliefs about the role of genes in disease processes, but not always accurately (Duncan, Parrott & Silk 2001). Many women sought testing for genetic contributors to breast cancer in the wake of media reports about breast cancer and genes when they had no family history to suggest the presence of a gene related to increased susceptibility (Hoskins, Stopfer, Calzone, Merajver, Rebbeck, Garber & Weber 1995).

News values contribute to the way that the science of health is communicated (Bell 1991). The "recency" value—occurrence of an event on the day, in the week, or month of the story—guides the selection of material, but the value placed on "facts and figures" give the news authority, and, as suggested by the realities of a health science vocabulary and the need for numeracy skills, news stories about medical research will have plenty of these to call on. It has also been noted that news outlets appear to value "negativity," a reference to the reality that controversy often generates interest in a story, and—once more—our own health science and numeracy skills will guide our ability to sort the conflict from the useful content. Space limitations likely contribute to science stories that appear in newspapers without definitions of terms or explanations of relationships among scientific concepts, contributing to science illiteracy. Science writers do, however, often include quotations from experts in their stories, averaging three such quotes per story (Conrad 1999), and supporting the value that news places on "attribution" (Bell 1991).

Science writers themselves have said that to improve the public under-standing of medical research news, they should clarify statistics and scien-tific terms, qualify stories about experiments by reporting general trends as they emerge in an area rather than specific studies, and educate the public about the incremental nature of science (Cooper & Yukimura 2002). These aims reflect efforts to address health science literacy deficits. Science writ-ers have also said that they recommend that they avoid stories that do not have proven human applications, such as rat research, because they believe these raise false hopes and promises (Cooper & Yukimura 2002). This form of self-censorship, however, limits our own opportunities for health and health care advocacy in promising areas.

In addition to the "news," we come to our awareness about science through media entertainment venues. In fact, a movie was the most fre-quent response in some of the research I have been involved with where nearly 500 participants were asked, "What is the FIRST media message or image that comes to your mind with the phrase, 'genes and health'?" ($n=467$). The 107 (22.81%) participants who identified a movie included 14 participants who provided a general response (e.g., "science fiction" mov-ies, a "cloning" movie) and 17 participants who named *Gattaca* (15.89%), with another 17 identifying *Jurassic Park* (Parrott, Volkman, Ghetian, Weiner, Raup-Krieger & Parrott 2008). *Gattaca* tells a story in which genetic engineering of humans is portrayed as commonplace, leading to the elimination of disease and physical impairment, and one's DNA playing a primary role in determining social class. In the movie, one character's birth takes place without the aid of biotechnology and leads to suffering and extreme genetic discrimination. Virus disease portrayals in entertain-ment media have been examined, with the movie, *Outbreak*, which por-trays a deadly virus, decreasing viewers' beliefs in their control over health and increasing their belief in chance outcomes of health (Bahk 2001). As noted earlier, research has shown that the reaction to use of the word *mutation* to describe something that is a core part of our physical being is generally negative. This may relate to entertainment media images in which "mutants" are vilified. Science fiction images and cartoons, such as "Teenage Mutant Ninja Turtles" possibly contribute to our responses as well. Health communication translational efforts have focused on strategi-cally harnessing the effects of entertainment venues on health beliefs and practices via education efforts, known as edu-tainment. These efforts have included efforts to address family planning and adoption (e.g., Vaughan & Rogers 2000), preventive behaviors in the contexts of HIV/AIDS (e.g., Beck 2004; Do & Kincaid 2006), breast cancer (e.g., Hether, Huang, Beck, Murphy & Valente 2008), and cornea donation (Bae 2008). *ER* and *Grey's Anatomy* introduced storylines in 2005 about breast cancer genetics (Het-her et al. 2008). In the *ER* storyline, a character with a family history of breast cancer and positive genetic tests for BRCA1 gene mutation learned about her increased risk for breast and ovarian cancer and made the decision to have a double mastectomy, but not oophorectomy, after physician encouragement and consultation about risk reduction strategies. In the *Grey's Anatomy* story, a character who tested positive for BRCA1 with a family history for breast cancer sought to have ovaries, uterus, and breasts removed, and met resistance from her spouse and even several doctors as she sought a second opinion

A total of 599 women participated in an online survey revealing that exposure to each BRCA1 gene mutation storyline was associated with greater knowledge about the BRCA1 gene mutation, greater behavioral intentions to undergo breast cancer screening, and greater self-reported behavior change (Hether et al. 2008). Respondents reported positive attitudes toward preventative surgery and getting a second opinion, but they also showed no significant attitude change about early breast cancer detection and no knowledge change about family history as a breast cancer risk. Overall knowledge was most improved via viewing both shows— seeming to reinforce both repetition and the reality that two sides to the decision-making influence were presented via viewing both shows (Hether et al. 2008). Viewing *ER* exerted a significant effect on intention to get tested for the BRCA gene mutation, while just viewing *Grey's Anatomy* did not; combined viewing had the strongest influence on testing intentions (Hether et al. 2008).

Health communication translational researchers are challenged in their edu-tainment collaboration efforts, as TV industry professionals view them as "slow"—a likely outcome linked to assuring that requirements for translational research have been met before proceeding to communicate about a health issue in an entertainment venue. This may also contribute to advertisers' desires to move ahead of the educational approach and advertise directly to consumers (Bouman 2002). Thus, in addition to the news and entertainment, advertising assumes a prominent role in "educating" us about science. Traditional commercial appeals, such as cost comparisons, accessibility, and convenience, comprise core efforts to promote products, activities, and other consumer goods related to health and health care. Additionally, direct to consumer (DTC) advertisements, allowed only in the U.S. and New Zealand, appear as *help-seeking ads* to educate and encourage consumers to consult with their physician; *reminder ads* that aim to aid recall of product names but provide limited information about a product's use; and *product specific ads* that aim to promote a product and provide information about its safety and efficacy.

DTC advertising blends the public relations, marketing, and advertising aims to bring the commercial realm into every aspect of health. The ads often use a great deal of "scientese"—"scientific jargon to create the impression of a sound foundation in science for claims, without substantive empirical evidence to support the jargon used" (Haard, Slater & Long

2004, p. 412). Scientese has been found to increase persuasiveness of the ad regardless of the consumer's educational background in science or involvement with an ad topic (Haard et al. 2004). These effects persisted whether there was a citation associated with the claims or not, a concern of FDA about the public's ability to assess evidence quality (Haard et al. 2004). DTC ads have been found to contribute to patients' requests for advertised antidepressants (Kravitz, Epstein, Feldman, Franz, Azari, Wilkes, Hinton & Franks 2005) and other prescription drugs (Hollon 1999). Advertisements for genetic testing have made their way to online sites (Williams-Jones 2003). These DTC ads have been criticized for lacking a social context surrounding genetics, presenting complex information, and ignoring the lack of consensus about the clinical utility of some tests (Gollust, Hull & Wilfond 2002). This finding harkens back to the first criterion to be applied before considering health communication translational efforts, whether commercial or otherwise. Some DTC ads seek to sell products, including genetic tests, which lack data about their accuracy and reliability. This makes their utility questionable (Wasson, Cook & Helzlsouer 2006). Product placement increases the strategic positioning of advertisements, often incorporating actual use of products and services into entertainment, advertising, and even the news without disclosure under the premise that it is a natural part of the work. Health policies have been adopted to limit the placement of "harmful" products, such as cigarettes, in traditional advertisements. Products deemed to be healthy, those without categorization as either harmful or healthy, and even some of those for which known harms exist make their way into influential venues.

Viewers may have no awareness of the products until viewing them in entertainment outlets, and the influence of viewing may not be subliminal, but it is often subtle (Galician 2004). In the *ER* and *Grey's Anatomy* episodes devoted to discussion of genetic contributors to breast cancer, for example, viewers gained awareness of genetic tests—a product linked to the effort to increase viewers' understanding about genetics and breast cancer; viewers subsequently increased intentions to seek such testing, regardless of their family history. This represents a challenge for health communication translational researchers in this era, as previously discussed, because family history should guide the choice to be tested for genetic contributors, or costs associated with testing will surpass economic gains associated with disease reduction linked to testing. The episodes emphasized the importance of a family history, but viewers demonstrated knowledge of the link between cancer and family history, while not extending it to their own situation regarding intent to be tested.

Product placement and DTC ads represent a particularly potent form of emotional influence, appearing, for example, in a playbill for a theatre presentation when a biotechnology company advertised its commercial test for the BRCA1 and BRCA2 genetic mutation in a play about a woman's painful death from ovarian cancer. The promotion was found to contain inaccurate information and did NOT promote contacting health care providers for considered discussions but instead promoted direct access to the product (Hull & Prasad 2001). Kaiser Permanente (Denver, CO) and Myriad Genetics Research conducted a DTC-ad campaign about their services. The number and pretest probability of referrals approved for cancer genetic services were compared with the previous year and a control condition using referral rates for the two years at the Henry Ford Health System in Detroit, MI (Mouchawar et al. 2005). At the Kaiser Permanente site, referrals increased 244% when comparing the two time periods; the Detroit site had no significant change between the two time periods (Mouchawar et al. 2005). The proportion of referrals for testing based on a high pretest probability of a mutation *dropped* from 69% in the year prior to the campaign to 48% during the campaign at the Kaiser Permanente site; there were no significant differences in referrals for the two years at the Detroit site (Mouchawar et al. 2005). This leads to the conclusion that DTC ads contributed to patients' increased requests for cancer genetic services and an increase in referrals based on patient requests rather than test results suggesting a need to have services.

CONCLUSION: A CAUTIONARY TALE

Having awareness of a genetic contributor to a health condition can be used to guide medical decisions (Jallinoja & Aro 2000). The science linking multiple genes to common conditions (Collins & McKusick 2001) represented in media and interpersonal forums guides us and our physicians to talk about causation and blame related to the presence of such genes. In an ideal world where health care costs were not the pendulum swinging on the metronome of decision-making, genetic counselors would advise patients about the implications of their decisions to be tested and public health communication would prepare us to understand the role of family health histories in an era of genomic health care. In the fiscally fractured world of the 21st century, counseling about these matters seems unlikely, while discrimination seems only too likely. Individuals may, for example, generalize their understanding about a genetic component of physical diversity between races to a genetic component of behavior, when culture may contribute more significantly to the latter (Braun 2002).

Awareness of genetic contributors to health can contribute to an overall level of attention to events and situations that may help prevent harmful health outcomes linked to our susceptibility for disease based on our genes. The medical and ethical effectiveness of communication about these issues depends fundamentally upon understanding how the lay public views the role of heredity for health. The discoveries naming multiple genes in relation to common health problems affect medical practice, as well as individual and society's expectations about responsibility and blame. One thing is certain: knowledge relating genes to health is fast outdistancing the expertise of patients and their health caregivers—a challenge that puts health communicators in the crossfire associated with communicating about genes and health.

REFERENCES

Andrykowski, M. A., Munn, R. K. & Studts, M. S. (1996). Interest in learning of personal genetic risk for cancer: A general population survey. *Preventive Medicine* 25: 527–536.

Bae, H. (2008). Entertainment-education and recruitment of cornea donors: The role of emotion and issue involvement. *Journal of Health Communication* 13: 20–36.

Bahk, C. M. (2001). Drench effects of media portrayal of fatal virus disease on health locus of control beliefs. *Health Communication* 13: 187–204.

Beck, V. (2004). Working with daytime and prime-time television shows in the United States to promote health. In A. Singhal, M. J. Cody, E. M. Rogers & M. Sabido (Eds.), *Entertainment-education and social change: History, research, and practice*. Mahwah, NJ: Erlbaum. 207–224

Bell, A. (1991). The language of news media. Cambridge, MA: Basil Blackwell.

Bouman, M. (2002). Turtles and peacocks: Collaborations in entertainment-education television. *Communication Theory* 12: 225–244.

Braun, L. (2002). Race, ethnicity, and health: Can genetics explain disparities? *Perspectives in Biology and Medicine* 45: 159–174.

Brodie, M., Foehr, U., Rideout, V., Baer, N., Miller, C., Flournoy, R. & Altman, D. (2001). Communicating health information through the entertainment media. *Health Affairs* 20(1): 192–199.

Burgess, M. M. (2001). Beyond consent: Ethical and social issues in genetic testing. *Nature Reviews/Genetics* 2: 147–151.

Burke, W., Pinsky, L. E. & Press, N. A. (2001). Categorizing genetic tests to identify their ethical, legal, and social implications. *American Journal of Medical Genetics* 105: 233–240.

Carnovale, B. V. & Clanton, M. S. (2002). Genetic testing: Issues related to privacy, employment and health insurance. *Cancer Practice* 10: 102–104.

Cavallari, L. H. & Limdi, N. A. (2009). Warfarin pharmacogenomics. *Current Opinion in Molecular Therapy* 11: 243–251.

Chase, G. A., Faden, R. R., Holtzman, N. A., Chwalow, A. J., Leonard, C. O., Lopes, C. & Quaid, K. (1986). Assessment of risk by pregnant women: Implications for genetic counseling and education. *Social Biology 33*: 57–64.

Clayton, E. W. (2003). Ethical, legal, and social implications of genomic medicine. *New England Journal of Medicine* 349: 562–569.

Collins, F. S., & McKusick, V. A. (2001). Implications of the human genome project for medical science. *Journal of the American Medical Association* 285: 540–544.

Condit, C. M., Dubriwny, T., Lynch, J., & Parrott, R. (2004). Lay people's understanding of and preference against the word "mutation." *American Journal of Medical Genetics* 130A: 245–250.

Condit, C. & Parrott, R., (2004). Perceived levels of health risk associated with linguistic descriptors and type of disease. *Science Communication* 26: 152–161.

Conrad, P. (1999). Uses of expertise: Sources, quotes, and voice in the reporting of genetics in the news. *Public Understanding of Science* 8: 285–302.

Cooper, C. P. & Yukimura, D. (2002). Science writers' reactions to a medical 'breakthrough' story. *Social Science & Medicine* 54: 1887–1896

Crowther, M. A. & Kelton, J., (2003). Congenital thrombophilic states associated with venous thrombosis: A qualitative overview and proposed classification system. *Annals of Internal Medicine* 138: 128–134.

Do, M. P. & Kincaid, L. (2006). Impact of an entertainment-education television drama on health knowledge and behavior in Bangladesh: An application of propensity score matching. *Journal of Health Communication* 11: 301–325.

Duncan, V., Parrott, R. & Silk, K. (2001). African American women's perceptions of the role of genetics in breast cancer risk. *American Journal of Health Studies* 17: 50–58.

Finkler, K. (2000). *Experiencing the new genetics: Family and kinship on the medical frontier.* Philadelphia: University of Pennsylvania Press.

Friede, A., Grossman, R., Hunt, R., Li, R. M. & Stern, S. (Eds.) (2003). National biospecimen network blueprint. Durham, NC: Constella Group, Inc.

Friedrich, M. J. (2004). Public education critical to population-wide genomics research. *Journal of the National Cancer Institute* 96: 1196–1197.

Galician, M. (2004). Introduction: Product placements in the mass media: Unholy marketing marriages or realistic story-telling portrayals, unethical advertising messages or useful communication practices? In M. Galician (ed.), *Handbook of product placement in the mass media.* Binghamton, NY: The Haworth Press, Inc. 1–5.

Genetic Information Nondiscrimination Act (GINA) of 2008. Accessed online at http://www.genome.gov/Pages/PolicyEthics/GeneticDiscrimination/GINAOmfpDpc/pdf. August 24, 2009.

Giorgianni, S. J. (1998). Communication: The essence of quality health care. *The Pfizer Journal* 2: 1.

Goldman, D., Oroszi, G. & Ducci, F. (2005). The genetics of addictions: Uncovering the genes. *Focus* 4: 401–415.

Gollust, S. E., Hull, S. C. & Wilfond, B. S. (2002). Limitations of direct-to-consumer advertising for clinical genetic testing. *Journal of the American Medical Association* 288: 1762–1767.

Grimes, D. A. & Snively, G. R. (1999). Patients' understanding of medical risks: Implications for genetic counseling. *Obstetrics & Gynecology* 93: 910–914.

Gurmankin, A. D., Domchek, S., Stopfer, J., Fels, C. & Armstrong, K. (2005). Patients' resistance to risk information in genetic counseling for BRCA1/2. *Archives of Internal Medicine* 165: 523–529.

Haard, J., Slater, M. & Long, M. (2004). Scientese and ambiguous citations in the selling of unproven medical treatments. *Health Communication* 16: 411–426.

Henderson, B. J. & Maguire, B. T. (2000). Three lay mental models of disease inheritance. *Social Science & Medicine* 50: 293–301.

Henderson, L. & Kitzinger, J. (1999). The human drama of enetics: 'Hard' and 'soft' media representations of inherited breast cancer. *Sociology of Health & Illness* 21: 560–578.

Hether, H. J., Huang, G. C., Beck, V., Murphy, S. T. & Valente, T. W. (2008). Entertainment-education in a media-saturated environment: Examining the impact of single and multiple exposures to breast cancer storylines on two popular medical dramas. *Journal of Health Communication* 13: 808–823.

Hirsh, J., Dalen, J. E., Anderson, D. R., Poller, L., Bussey, H., Ansell, J. & Deykin, D. (2001). Oral anticoagulants: Mechanism of action, clinical effectiveness, and optimal therapeutic range. *Chest* 119: 8S–21S.

Hollon, M. F. (1999). Direct-to-consumer marketing of prescription drugs: Creating consumer demand. *Journal of the American Medical Association* 281: 382–384.

Hoskins, K. F., Stopfer, J. E., Calzone, K. A., Merajver, S. D., Rebbeck, T. R., Garber, J. E. & Weber, B. L. (1995). Assessment and counseling for women with a family history of breast cancer: A guide for clinicians. *Journal of the American Medical Association* 273: 577–585.

Hull, S. C. & Prasad, K. (2001). Reading between the lines: Direct-to-consumer advertising of genetic testing in the USA. *Reproductive Health Matters* 9: 44–48.

Jallinoja, P. & Aro, A. R. (2000). Does knowledge make a difference? The association between knowledge about genes and attitudes toward gene tests? *Journal of Health Communication* 5: 29–39.

Juengst, E. T. (1999). Genetic testing and the moral dynamics of family life. *Public Understanding of Science* 8: 193–205.

Kelly, K., Love, M., Pearce, K., Porter, K., Barron, M. & Andrykowski, M. (2009, in press). Physician identification of hereditary and familial cancers in rural Appalachia. *Journal of Rural Health* 25: 372–377.

Kravitz, R. L., Epstein, R. M., Feldman, M. D., Franz, C. E., Azari, R., Wilkes, M. S., Hinton, L. & Franks, P. (2005). Influence of patients' requests for direct-to-consumer advertised antidepressants: A randomized controlled trial. *Journal of the American Medical Association* 293: 1995–2002.

Lerman, C., Shields, P. G., Wileyto, E. P., Audrain, J., Pinto, A., Hawk, L., Krishnan, S., Niaura, R. & Epstein, L. (2002). Pharmacogenetic investigation of smoking cessation. *Pharmacogenetics* 12: 627–634.

Marteau, T., Senior, V., Humphries, S. E., Bobrow, M., Cranston, T., Crook, M. A., Day, L. Fernandez, M., Home, R., Iversen, A., Jackson, Z., Lynas, J., Middleton-Price, J., Savine, R., Sikorski, J., Watson, M., Weinman, J., Wierzbicki, A. S. & Wray, R. (2004). Psychological impact of genetic testing for familial hypercholesterolemia within a previously aware population: A randomized controlled trial. *American Journal of Medical Genetics Part A* 128A: 285–293.

Mouchawar, J., Hensley-Alford, S., Laurion, S., Ellis, J., Kulchak-Rahm, A., Finucane, M., Meenan, R., et al. (2005). Impact of direct-to-consumer advertising for hereditary breast cancer testing on genetic services at a managed care organization: A naturally-occurring experiment. *Genetics in Medicine* 7: 191–197.

Moyer, A., Greener, S., Beauvais, J. & Salovey, P. (1995). Acuracy of health research reported in the popular press: Breast cancer and mammography. *Health Communication* 7: 147–161.

Murff, H., Byrne, D. & Syngal, S. (2004). Cancer risk assessment: Quality and impact of the family history interview. *American Journal of Preventive Medicine* 27: 239–245.

Parrott, R. (2008). A multiple discourse approach to health communication: Translational research and ethical practice. *Journal of Applied Communication Research* 36: 1–7.

Parrott, R., Silk, K., Weiner, J., Condit, C., Harris, T. & Bernhardt, J. (2004). Deriving lay models of uncertainty about genes' role in illness causation to guide communication about human genetics. *Journal of Communication* 54(1): 105–122.

Parrott, R., Silk, K., Dillow, M., Krieger, J., Harris, T. & Condit, T. (2005). The development and validation of tools to assess perceptions of genetic discrimination and genetic racism. *Journal of the National Medical Association* 97: 980–991.

Parrott, R., Volkman, J., Ghetian, C., Weiner, J., Raup-Krieger, J. & Parrott, J. (2008). Memorable messages about genes and health: Implications for direct-to-consumer marketing of genetics tests and therapies. *Health Marketing Quarterly* 25(1): 1–25.

Petrucelli, N., Walker, M. & Schorry, E. (1998). Continuation of pregnancy following the diagnosis of a fetal sex chromosome abnormality: A study of parents' counseling needs and experiences. *Journal of Genetic Counseling* 7: 401–415.

Rodgers, J. E. (1999). Overarching considerations in risk communication: Romancing the message. *Journal of the National Cancer Institute* 25: 21–22.

Sanderson, S. C. & Wardle, J. (2005). Will genetic testing for complex diseases increase motivation to quit smoking?: Anticipated reactions in a survey of smokers. *Health Education & Behavior* 32: 640–653.

Schwartz, L. M., Woloshin, S., Black, W. C. & Welch, G. (1997). The role of numeracy in understanding the benefit of screening mammography. *Annals of Internal Medicine* 127: 1013–1022.

Seppa, N. (2001). Gene therapy for sickle-cell disease? *Science News 160*: 372.

Vaughan, P. W. & Rogers, E. M. (2000). A staged model of communication effects: Evidence from an entertainment-education radio soap opera in Tanzania. *Journal of Health Communication* 5: 203–227.

Wang, C., Gonzalez, R., Milliron, K. J., Strecher, V. J. & Merajver, S. D. (2005). Genetic counseling for BRCA1/2: A randomized controlled trials of two strategies to facilitate the education and counseling process. *American Journal of Medical Genetics* 134A: 66–73.

Wasson, K., Cook, E. D. & Helzlsouer, K. (2006). Direct-to-consumer online genetic testing and the four principles: An analysis of the ethical issues. *Ethics & Medicine* 22: 83–91.

Weiner, J. L., Silk, K. J. & Parrott, R. L. (2005). Family communication and genetic health: A research note. *The Journal of Family Communication* 5: 313–324.

Williams-Jones, B. (2003). Where there's a Web, there's a way: Commercial genetic testing and the Internet. *Community Genetics* 6: 46–57.

Zillmann, D. & Brosius, H. (2000). *Exemplification in communication: The influence of case reports on the perceptions of issues.* Mahwah, NJ: Lawrence Erlbaum Associates.

Part III

The Media, the Public, and the Person

8 Media Misinformation and the Obesity Epidemic

The Conflict Between Scientific Fact and Industry Claims

Steven Giles and Marina Krcmar

As 401(k)s and Wall Street's fat cats slim down, there is no denying that the waist size of Americans continues to hit record highs. Recent data compiled by the Centers for Disease Control's Behavioral Risk Factor Surveillance System (BRFSS) indicates that in 1991 only four states had obesity prevalence rates of 15–19% and no states exceeded 20%. By 2004, however, seven states had obesity rates of 15–19%, 33 states had rates of 20–24%, and nine states had rates that exceeded 25%. In fact, by 2000 "the human race reached a sort of historical landmark, when for the first time in human evolution the number of adults with excess weight surpassed the number of those who were underweight" (Caballero 2007). This has led many researchers to conclude that we are witnessing a domestic as well as international obesity epidemic (Flegal, Carroll, Ogden & Johnson 2002; Mokdad et al. 2003; Ogden et al. 2006). Indeed, President Obama acknowledged the problem in his January 27, 2010 State of the Union Address when he praised his wife for "creating a national movement to tackle the epidemic of childhood obesity and make our kids healthier" (Obama, 2010).

With the problem of obesity we are, of course, speaking about a matter of "public health." Therefore, it is important that the public be informed and instructed on how to deal with its presence and consequences. Communication and rhetorical competence on the part of health care professionals is thereby called upon to address this challenge. When this competence is lacking or even when it is being capitalized on to hide or exacerbate the problem—as, for example, in media advertisements—the issue at hand warrants even more careful attention. Clearly, then, the media's coverage of the issue of obesity is not only a matter of public health but one of ethics as well.

Within the field of medicine, the combined study of public health and communication is typically associated with teaching and research in the area of bioethics. Social scientists and rhetorical scholars from the discipline of communication studies also emphasize the ethical and discursive complexities involved in matters of public health (Hyde 2008; Hyde & King 2010). In his recent book, *The Future of Bioethics,* Howard Brody (2009) makes much of such "interdisciplinary" efforts for establishing the legitimacy of "bioethics as an intellectual pursuit" (pp. 3–48). The present essay is offered to facilitate this pursuit. Specifically, we will discuss how media exposure in general, and food advertisements in particular, have played a central role in the rise of obesity in the U.S.

We will defend this thesis by adopting a seminal piece by Bushman and Anderson (2001) in which they compare the contentious social debate concerning the effect of television violence on aggression with an earlier, and equally heated debate about the effect of cigarette smoking on lung cancer. With a carefully crafted set of arguments they describe how the scientific community knew of the link between smoking and lung cancer long before it was understood by laypersons. They argued that the tobacco industry undermined evidence supporting the link between smoking and cancer in order to sustain and increase their own profits; ultimately only through government intervention did the tobacco industry begin to take some measure of responsibility for the effects of their product. Similarly, the authors argue, the link between media violence exposure and aggression is well-accepted by a majority of the research community but poorly understood by laypersons. The media industry has a vested interest in minimizing and undercutting evidence supporting the link between exposure to media violence and aggression, and has therefore actively sought to argue that no conclusive evidence exists. Bushman and Anderson's (2001) final contention is that researchers themselves are ethically responsible for engaging not only in the scholarly quest of evidence gathering but also in the practical pursuit of activism. We need to inform the public and actively publicize the results of academic research on the effects of and risks associated with, for example, media exposure.

Utilizing Bushman and Anderson's arguments, we will demonstrate that the relationship between television exposure, especially to food advertising on one hand, and obesity on the other, also follows a similar trajectory. As the scientific evidence linking exposure to food advertising and obesity mounts, the media have presented weight maintenance as an issue of individual personal responsibility. Through both direct arguments and through media portrayals, media have attempted to minimize the very real link between exposure to food advertising and obesity.

CHILDREN AND OBESITY

Children are particularly vulnerable to the obesity epidemic. The prevalence of obesity has doubled for children aged 2–5 years over the past three decades (Institute of Medicine). The National Health and Nutrition Survey found that nearly 21% of children ages two to five and 30% of children ages six to 19 are overweight or at risk for becoming overweight (Ogden, Flegal, Carroll & Johnson, 2002). The health risks of excess weight and obesity in childhood are well-established and include risk factors for cardiovascular disease, the onset of Type 2 diabetes, hypertension, hyperlipidemia, sleep apnea, and gall bladder disease (Dietz & Robinson 2005). Moreover, overweight children are significantly more likely to become overweight adults: 26%–41% of adults who are overweight were obese in preschool and 42%–63% were obese in elementary school (Serdula et al. 1993). In fact, some have argued that obesity in adulthood is perhaps the most serious consequence of childhood obesity, because it is linked with increased mortality and morbidity (Freedman, Khan, Dietz, Srinivasan & Berenson 2001; Lobstein, Baur & Uauy 2004).

ENVIRONMENTAL CAUSES OF OBESITY

Demonstrating that obesity is a social epidemic is not difficult. Agreeing on a set of causal factors, however, is more problematic. For instance, Keith and colleagues (2006) argue that it is generally accepted that the primary causes of obesity are reduced physical activity, due to increases in media use and reduced physical education in schools; and specific food and marketing practices such as increased portion size, availability of fast food, vending machines in schools, and the use of high fructose corn syrup. Indeed, the Kaiser Family Foundation report suggests that children's use of media, including computers, Internet, video games, and television, "may be one of the primary contributors to the poor fitness and obesity of many of today's adolescents" (Larwin & Larwin 2008, 938). But as Keith et al. note, each of these factors have a "less-than-unequivocal evidential basis" when one considers the effect sizes from empirical data. Keith et al. (1586) go on to claim that 10 other factors play just as great of a causal role as the media exposure and physical activity: sleep, debt, pollution, air conditioning, medication, population age/ethnicity, generational environmental change, older mothers, fertility, and obese couples. Their conclusion is that "Undue attention has been devoted to reduced physical activity and food marketing practices as postulated causes for increases in the prevalence of obesity, leading to neglect of other plausible mechanisms and well-intentioned, but potentially ill-founded proposals for reducing obesity rates" (1585).

Not surprisingly, a number of interventions have been designed to increase physical activity and reduce media use. These studies, however, often suffer from poor methodological designs (Zoeller 2009). For instance, Larwin and Larwin (2008) employed a single-subject design to test the potential to increase physical activity and decrease media use in a 14-year-old female adolescent. Even in well-designed studies, such as a randomized controlled trial of the Pathways program (Caballero et al. 2003), there was no effect found for program participation on BMI. Thus, while there is evidence to suggest that physical activity and general media use are associated with weight status (Marshall, Biddle, Gorely, Cameron & Murdey 2004; Zoeller 2009), there is limited evidence to support the effectiveness of exercise and media use interventions in reducing obesity (Keith et al. 2006). The evidence to support the effectiveness of commercially available self-help weight loss programs is also negligible. For instance, in their review of randomized controlled trials of commercially available weight loss programs, Tsai and Wadden (2005) found that the greatest weight loss occurred in one investigation of Weight Watchers, where participants had lost 3.2% of their original weight after two years. But when attrition is considered, the results of this study as well as studies in clinical medical settings is less than optimal.

What can explain this problem? Why are weight loss interventions that encourage less media use and more physical activity ultimately unsuccessful? Perhaps the problem lies in the primary focus of interventions and their assumption that the people responsible for the obesity epidemic are the overweight individuals themselves. In other words, perhaps the assumption is inaccurate that obesity can be controlled if individuals' learn to curb their appetites and increase their level of physical activity. These interventions place the responsibility solely on the individual. However, the exponential rise in obesity over the past 30 years (Keith et al. 2006), suggests that the increase in obesity is more a consequence of social environmental influences than personal will power or choice. Simply put, it is unlikely that human beings have substantially changed in regards to personal efficacy and health literacy in the past 30 years. After all, one need only consider the average expenditures on diet products, sales of fashion magazines, and rates of disordered eating to conclude that Americans do indeed care about weight and appearance. Moreover, the gene pool has not significantly changed (Mokdad et al. 2000). Thus, it seems safe to say that the rise in obesity levels is not due to changes in the individual but changes in the social environment. This is not to say that individuals do not bear any responsibility for their health or weight status. Rather, we argue that a more productive view of the problem of obesity would be to conceptualize the epidemic as being ecological in nature. A broader, social environmental approach to the obesity epidemic and to weight loss allows for large-scale social changes that are more likely to impact larger numbers of people.

For example, social ecological theory (Cohen, Scribner & Farley 2000) argues that past interventions have focused on the problem behaviors themselves (e.g., excessive food consumption, binge eating). Instead, a more beneficial approach would be to enhance health outcomes by manipulating or altering environments and the conditions under which people enact problem behaviors. Interventions, therefore, whether programs, campaigns, or legislation, should focus on these broader ecological and social contexts. Cohen goes on to suggest that behavior—in this case those behaviors that lead to obesity—are themselves caused by four categories of structural factors, each of which can be addressed environmentally, socially, and politically in order to encourage positive health outcomes. First, for a problem behavior to occur, the harmful consumer product must be readily available. In the case of some harmful products, such as cocaine, laws are in place to prohibit its possession, thus reducing its availability. Less severe restrictions, such as the age restrictions in place for alcohol and tobacco, can also curtail possession. However, other consumer products that are potentially harmful, such as calorie dense and innutritious foods, are readily available. Second, Cohen discusses the physical characteristics of products, including their physical structure, their design, and their quality. In the case of food, ecological theory would suggest that foods do not need to be made with saturated fat, growth hormones or any number of unhealthful characteristics. In fact, the European Union legislates against problematic food characteristics such as pesticides or vegetables grown with growth hormones (Jukes, n.d.). Third, ecological theory states that social structures such as laws and policies can encourage or prohibit behavior. For example, seatbelts were readily available in motor vehicles since 1958 but were not used consistently by passengers. Currently, however, with the exception of New Hampshire, all states have a mandatory seat belt law. Seatbelt use subsequently increased significantly after passage of the law in these states (Governors Highway Safety Association, 2011). In this case, social legislation was extraordinarily successful when campaigns were not (or were to a much lesser degree).

Fourth, ecological theory states that media and cultural messages have a strong influence on behavior and can encourage some behaviors (e.g., physical activity) and constrain others by changing individual level attitudes and beliefs. In fact, past research suggests that media do, in fact, have an effect on behavioral intentions and outcomes, particularly in children and adolescents. For example, a large body of evidence suggests that violent media influences the aggression and moral reasoning of young children (Krcmar & Farrar 2009; Krcmar & Lachlan 2009). Similarly, exposure to thin-depicting media is related to social norms regarding thinness and disordered eating intentions among college students (Giles, Helme & Krcmar 2007; M. Krcmar, Helme & Giles 2008).

Overall, then, the first three structural features can directly influence individuals through facilitating or constraining behavior. The fourth, media, operates by changing variables including attitudes, cognitions, and social norms. Given the factors involved in any given behavior, and given our earlier argument that individual will power and genetic makeup are unlikely to have changed in the last 30 years, it makes sense that social ecology best explains the rise in obesity. Social ecology argues that while the individual is an active agent in the decision-making process, one's health behavior is a function of the intersection of intrapersonal, interpersonal, organizational, community, and public policy factors that ultimately play into attitudes and motivations (Sallis, Owen & Fisher 2008). Finally, it is these more distal factors that can offer a cogent explanation for obesity rather than the more simplistic explanation that cites overeating and sedentary lifestyles. While food consumption and exercise are clearly related to obesity, these proximal variables do not provide a very thorough explanatory mechanism nor do they offer an entrée into a more complex understanding that is necessary to best interpret and counteract an epidemic.

In the following sections, we utilize early arguments made by the tobacco industry to undercut links between smoking and lung cancer as a template for the current debate about media and obesity. In each section we start by presenting an argument made by the tobacco industry (e.g., Not everyone who smokes gets lung cancer) and we then present a similar argument that has been made to undercut the link between media exposure and obesity. Specifically, we discuss how media have made their case, either directly or indirectly, and we offer counter-evidence, supporting the argument that media exposure, especially to food advertising, is related to obesity. We present three instructive parallels between the smoking and lung cancer relationship and the media and obesity debate and then provide evidence from research on media effects that helps to explain why consumers may be unaware of the harmful effects of media advertising.

DENYING THE EVIDENCE:
THE TOBACCO INDUSTRY LEADS THE WAY

(1) Mere Exposure is Not a Sufficient Explanation for Causal Effects

The first way in which the tobacco industry sought to deflect claims that smoking caused cancer was by pointing out that exposure to tobacco could not cause cancer because so many other factors were related to it. Similarly, the advertising industry can claim that there are a number of other factors that influence the development of obesity, including sedentary lifestyles, and greater use of convenience foods (Hoek & Gendall 2006; Lvovich 2003). In other words, since the media's contribution to obesity is small in comparison to other contributing factors, then food advertising cannot be viewed as a sufficient explanation for childhood obesity. Advertisers further argue that food advertisements are not designed to change demand for a food category (e.g., cereal) but aim to change brand preference (Young 2003). Finally, even if these food ads could be proven to shape food choices of children, it is the responsibility of parents to make those food choices and moderate their child's exposure to media. In other words, given the complex nature of obesity it is impossible to lay blame on companies that advertise unhealthy food to children. These arguments serve to deflect any responsibility the food industry has in marketing to children and to shift the blame to parents.

It is true that a number of factors contribute to the obesity epidemic. But is this a sufficient reason to disregard the influence of food promotions on children's nutrition? First, we are not talking about "mere" exposure to advertisements.

Children view over 40,000 commercials per year and most of these commercials advertise fast food, toys, candy, and cereal (Kunkel & Gantz 1992). Harrison and Marske (2005) content analyzed the nutrition labels of 275 foods that were advertised during television shows that are heavily watched by children. They found that 83% of advertised foods were for convenience/fast foods and sweets and that a child who would eat according to the diet depicted in the ads would exceed the recommended daily values for total fat, saturated fat, and sodium. More recently, Powell et al. (2007) content analyzed over 224,000 advertisements that aired on the top 170 channels viewed by children between two and 11 years of age. Food advertisements comprised over one-quarter of non-program content and over one-third of all advertisements that aired during or between television programs. Advertisements for cereal were most common, comprising 27.4% of all food content advertised. The authors conclude that there is evidence of a recent upward trend between 2000–2005 in how much exposure children have to food advertisements.

Second, food advertisements do in fact shape brand preference. For instance, Connor (2006) found that more than half of all food ads in her content analysis were aimed at children, with the majority of those ads being for fast food or sweetened cereal. The fast food ads, in particular, utilized strategies to build positive associations with the product, such as licensed characters and logos. Connor argues that these strategies aim to do more than impact immediate purchasing behavior; they are designed to build brand loyalty.

With that being said, food advertisements go beyond shaping brand preference; they influence children's understanding of nutrition and preference for certain food categories. In their report to the World Health Organization, Hastings and colleagues (2006) reviewed over 38 studies that met at least one criterion for causality. They concluded that food promotion has little effect on what children view as healthy, but within certain contexts it can influence nutritional knowledge. That is, generally speaking, exposure to food advertisements for low nutritional food was associated with poorer nutritional knowledge among children. Moreover, the authors reviewed 14 experimental studies that found strong causal evidence that exposure to food advertisements was associated with significant changes in children's food preferences. Children who were exposed to food advertisements were more likely to prefer high fat, salt, or sugar foods compared to those not exposed to the food advertisements. More recently, Dixon and her colleagues (2007) found that heavier television use and greater exposure to commercial television were positively associated with greater junk food consumption among 5th and 6th grade students. This evidence undermines the advertising industry's argument that food ads are designed to change brand preference and not to shape food category preferences.

Lastly, while parents should provide oversight of media use among their children, research indicates that kids' time spent watching television is associated with requests to parents for advertised foods (Taras, Sallis, Patterson, Nader & Nelson 1989). Even experimental studies reveal that exposure to advertising affects children's food preferences and their efforts to shape the foods their parents buy (Coon, Goldberg, Rogers & Tucker 2001; Hitchings & Moynihan 1998).

(2) Denial, Denial, and More Denial

A second way that the tobacco industry dodged responsibility was by denying any association b etween tobacco use and heart disease while making the tobacco debate one of personal freedom. Much like the early response of the tobacco industry, media, in its portrayal of weight loss and obesity, has downplayed and in some instances denied the link between food advertising exposure and obesity. Instead, they focus on the role of personal responsibility in weight management and ignore the social psychological ramifications of environmental issues, such as media exposure. Obesity is framed as a personal problem, to be overcome with will power and exercise; advertising is ignored as a meaningful potential source of effects. Consider the recent framing of obesity: television programs like *The Biggest Loser* showcase several people who are, in most cases, morbidly obese. The program then documents their trials and challenges as they battle food and attempt to exercise. While programs such as this celebrate human will power to overcome problems, they ignore issues about cultural and environmental factors that may play a causal role in obesity. The program, for example, pointedly does not showcase the long trajectory of obesity onset. This trajectory often includes early norming of unhealthy diet habits, created in part by food advertising. Instead, obesity is framed as a personal problem with an individual solution. That framing certainly makes for better programming, even as it ignores the effects of cultural and environmental factors on obesity.

In fact, in a recent national survey of more than 2,500 adults, fully 83% of adults believe that obesity, even in children, can be blamed on parents. Only 27% blamed food and beverage companies whereas 11% blamed media (Bright 2007). Although there is a rising concern among parents that both the food and beverage companies as well as the media who advertise food products is partly responsible for obesity, the industry itself continues to downplay the effects. For instance, the Center for Consumer Freedom, a non-profit organization that receives support from resturants and food companies has asserted that "life-style, not diet, is the main causeof obesity" (The Center for Consumer Freedom, n. d.)

Systematic research has also examined the media framing of the obesity debate. Lawrence (2004) examined newspaper articles from 1985–2003 that addressed the issue of obesity. Whereas the earliest articles were likely to frame obesity as a personal and behavioral issue, changes in the framing began to emerge in the mid-1990s. As news articles began to frame the issue as one of environmental and social concern (including the effects of the food and beverage industry), however, something of a backlash occurred. At the publication of the research, Lawrence claimed that the "struggle is far from over." Framing of the obesity issue had once again introduced the notion of personal responsibility. Industry was seen as a small, and somewhat insignificant part of the problem. How is it that we remain duped? How is it that audiences and legislators alike are willing to accept that the food industry does not bear significant burden for the outcomes of the advertised products?

(3) Severe, Long-term Effects Appear Unlikely Given that Short-term Effects are Minimal

A third parallel exists between the smoking debate and the current debate linking food ads to obesity. The long-term cumulative effects of smoking are much more severe than one, single exposure to a cigarette might be. Thus, demonstrating effects requires extensive longitudinal research, rather than, and in addition to, single-shot experiments. Similarly, whereas exposure to a single advertised food product is likely to have only a small and short-term impact on food preferences (Dixon et al. 2007), repeated exposure is likely to create life-long food schema and attitudes toward food that shape eating patterns. It is these attitudes and behavioral schema that are likely to contribute to obesity, thus making the link between exposure to food advertisement and obesity more difficult to detect.

Specifically, research has demonstrated that the link between short-term, single-shot exposure to food advertising can demonstrate that advertising increases appetite and positive attitudes toward advertised foods, especially in children (Dixon et al. 2007). Experiments are subject to the criticism that increasing positive attitudes towards an advertised food in the short-term does not allow us to conclude that food advertising is causally related to obesity. Therefore, researchers have undertaken additional work, outside of the laboratory setting, in order to demonstrate a causal link between food advertising exposure and obesity. For example, one large-scale study found that African American teens and adults not only make up a proportionately higher number of those suffering from obesity, Type 2 diabetes and hypertension, but generally, they watch more television and the television they do watch contains more advertisements for food products, especially fast food, candy, sweetened beverages and junk food (Powell, Szczypka & Chaloupka 2007). In addition, in a diary study of parents and children between the ages of two and 12, researchers found that children's exposure to food advertising was significantly related to their consumption of advertised brands and energy-dense product categories. The relation between advertising exposure and overall food consumption held especially in lower-income families (Buijzen, Schuurman & Bomhof 2008). Similar studies, using longitudinal survey data, have found identical results: greater time spent with television and greater exposure to food advertising was linked to more positive attitudes towards junk food and fast food and to greater consumption of these food products (Dixon et al. 2007). Thus, the evidence emerging from survey research, diary studies, and longitudinal data suggest that there is some long-term link between exposure to food advertising and positive attitudes toward and consumption of high calorie foods.

THE THIRD-PERSON EFFECT

Given the longitudinal data and convergence of evidence that point to the negative effects of food advertising on consumer behavior, we might expect consumers to demand greater responsibilty on the part of advertisers. But that has not been the case, in part because defenders of the food industry have confounded the debate by questioning the science that implicates advertising and making restrictions on advertising an issue of personal freedom, and in part because of a well-known phenomenon called the "third person effect" The third-person effect is the phenomenon by which individuals tend to believe that others are more susceptible to the effects of stimuli (e.g., smoking, violent media, food advertising) than they, themselves are (Salwen 1998). This placates individuals into consuming the product while the industry continues to peddle it, despite the mounting evidence against the product.

Consider, for example, food portrayals in the media and food advertising. There remains little doubt that food advertising is common on programs targeting both adults and children. In particular, food ads on programs targeting children and adolescents make up a large proportion of the television landscape. Food is among one of the top two products advertised to children, with the other being toys (Williams, Achterberg & Sylvester 1993). Furthermore, food advertising most typically advertises unhealthy foods. According to the Kaiser Family Foundation report, released in 2007 (Program for the Study of Media and Health 2007), the most frequent foods advertised are fast food, snack foods, candy, and sugared cereals. Thus, repeat exposure to food advertising is common among those exposed to television, and in particular, it is common among children and adolescents.

An interesting phenomenon occurs when individuals consume substances, whether those substances are cigarettes or food advertising. Whereas initial exposure to a particular stimulus may cause a heightened awareness of the stimulus or even generate a negative response, additional exposure tends to dampen affective responses. In the case of cigarette smoking, smokers often report nausea and dizziness initially but repeated exposure desensitizes the consumer to the negative physical and affective response. Similarly, initial exposure to food advertising may cause a heightened awareness to the hypocrisy of the ads. Audiences are shown high calorie foods being consumed by thin people. However, over time the repetitious nature of the stimuli dulls our responses to it. Desensitization takes place. We cease to be aware of any real affective responses and in fact, we may cease to notice the ads and product placements, themselves. Thus, long-term and repeated exposure, whether it be to media violence, tobacco smoke, or food advertising, tends to desensitize us to the stimuli.

In other words, in small doses, advertising catches our attention. Quick cuts, zooms, and music are likely to cause an increase in attention (Greer, Potts, Wright & Huston 1982). With increased exposure, however, awareness of advertising decreases through the process of desensitization. After desensitization occurs, individuals may well be aware of the negative impact of, say, smoking or advertising but through the process of personal desensitization, they may feel immune to these effects themselves. Many individuals believe that there is an effect of food advertising on food consumption. After all, a multibillion dollar industry like advertising must be able to demonstrate its effectiveness or it would not survive. However, most individuals also believe that food advertising has no effect on them, in part because they are unaware of how much food advertising and product placement they are exposed to. Thus, individuals demonstrate a robust third-person effect (David & Johnson 1998) due in part to their own desensitization to it.

PUTTING RESPONSIBILITY BACK INTO THE DEBATE

How might all of this relate to the role of, say, the fast food industry and its advertisers in ethical matters? Food advertising affects us and is indeed linked to obesity in children (Desmond & Carveth 2007). However, industry executives are quick to point out that individuals freely choose to eat (or avoid) the foods they provide. Thus, neither they nor their advertisements are ethically responsible for the health risks involved. Similarly, laypersons, who are prone to the third-person effect, believe themselves to be unaffected by food advertising. Together, then, food marketers and food consumers are complicit. Both believe, or want to believe that there is no effect of food advertising on food consumption, leaving food producers ethically free of responsibility for any negative health outcomes of fast food and junk food consumption.

In an effort to undermine the research, industry officials have long been capable of questioning research methods. For example, as discussed above, single-shot experiments are often criticized for demonstrating small effect-sizes and for being conducted in non-naturalistic environments. On the other hand, large-scale survey research and longitudinal studies are then criticized for lacking control and for not accounting for all possible factors that may cause effects such as cancer, on one hand, or obesity, on the other. Ultimately, individual studies are criticized while the relative consistency of the outcomes of various studies is ignored.

Cultural values regarding personal freedom and choice and a free market society may also work against ecological solutions that call for increased restrictions on junk food advertising. For many years the smoking lobby convinced Americans that smoking was a personal choice and any attempt by government to curb availability of cigarettes or to limit advertising was simply un-American. Evidence that elucidated the addictive nature of cigarettes, as well as research on the effects of secondhand smoking arguably shifted the debate away from personal choice and onto the responsibility of the industry for putting profit ahead of people. While evidence of the addictive qualities of food is still mounting, it is difficult for people to understand the consequences of obesity on society as a whole. Perhaps until the equivalent of secondhand exposure to obesity becomes a reality, it will be too easy for Americans to believe that obesity is largely an individual problem with individual-based solutions. It is incumbent upon the academic community, then, to better translate and publicize research findings through channels that will reach the general public.

REFERENCES

Bright, B. (2007). The Wall Street Journal: WSJ Online/Harris Interavtice Health-Care Poll. Polls Growing Concern About Role of Advertising in Child Obesity. Available from <http://online.wsj.com/article/SB18790b29508900233.html>

Brody, H. (2009). *The future of bioethics.* New York: Oxford University Press.

Brown, J. & Einsiedel, E. (1990). Public health campaigns: Mass media strategies. In E. Ray & L. Donohew (Eds.), *Communication and health: Systems and application.* Hillsdale, NJ: Lawrence Erlbaum. 153–169.

Buijzen, M., Schuurman, J. & Bomhof, E. (2008). Associations between children's television advertising exposure and their food consumption patterns: A household diary-survey study. *Appetite* 50(2–3): 231–239.

Bushman, B. & Anderson, C. (2001). Media violence and the American public: Scientific facts versus media misinformation. *American Psychologist* 56 (June/July): 477–489.

Caballero, B. (2007). The global epidemic of obesity: An overview. *Epidemiologic Reviews.*

Caballero, B., Clay, T., Davis, S. M., Ethelbah, B., Rock, B. H., Lohman, T., et al. (2003). Pathways: A school-based, randomized controlled trial for the prevention of obesity in American Indian schoolchildren. *American Journal of Clinical Nutrition* 78(5): 1030–1038.

Cohen, D. A., Scribner, R. A. & Farley, T. A. (2000). A structural model of health behavior: A pragmatic approach to explain and influence health behaviors at the population level. *Preventive Medicine* 30(2): 146–154.

Connor, S. M. (2006). Food-related advertising on preschool television: Building brand recognition in young viewers. *Pediatrics* 118(4): 1478–1485.

Coon, K. A., Goldberg, J., Rogers, B. L. & Tucker, K. L. (2001). Relationships between use of television during meals and children's food consumption patterns. *Pediatrics* 107(1), e7. doi. 10.1542/peds. 107.1.e7.

David, P. & Johnson, M. A. (1998). The role of self in third-person effects about body image. *The Journal of Communication* 48(4): 37–58.

Desmond, R. & Carveth, R. (2007). The effects of advertising on children and adolescents: A Meta-analysis. In R. Preiss, B. Gayle, M. Burrell, M. Allan & J. Bryant (Eds.), *Mass media effects research: Advances through meta-analysis.* Mahwah, NJ: Lawrence Erlbaum Associates. 169–180.

Dietz, W. H. & Robinson, T. N. (2005). Overweight children and adolescents. *New England Journal of Medicine* 352(20): 2100–2109.

Dixon, H. G., Scully, M. L., Wakefield, M. A., White, V. M. & Crawford, D. A. (2007). The effects of television advertisements for junk food versus nutritious food on children's food attitudes and preferences. *Social Science & Medicine* 65(7): 1311–1323.

Flegal, K. M., Carroll, M. D., Ogden, C. L. & Johnson, C. L. (2002). Prevalence and trends in obesity among US adults, 1999–2000. *Journal of the American Medical Association*: 288(14): 1723–1727.

Freedman, D. S., Khan, L. K., Dietz, W. H., Srinivasan, S. R. & Berenson, G. S. (2001). Relationship of childhood obesity to coronary heart disease risk factors in adulthood: The Bogalusa heart study. *Pediatrics* 108(3): 712–718.

Frosch, D. L., Shoptaw, S., Nahom, D. Jarvik, M. E. (2000). Associations between tobacco smoking and illicit drug use among methadone-maintained opiate-dependent individuals. *Experimental and Clinical Psychopharmacology* 8(1): 97–103.

Giles, S. M., Helme, D. & Krcmar, M. (2007). Predicting disordered eating intentions among incoming college freshmen: An analysis of social norms and body esteem. *Communication Studies* 58(4): 1–16.

Governors Highway Association the States Voice on Highway Safety. Seat Belt Laws. June, 2011. Available from <http://www.ghsa.org/html/stateinfo/laws/seatbelt_laws.html

Greer, D., Potts, R., Wright, J. C. & Huston, A. C. (1982). The effects of television commercial form and commercial placement on children's social behavior and attention. *Child Development* 53(3): 611–619.

Harrison, K. & Marke, A.L. (2005). Nutritional content of foods advertised during the television programs children watch most. *American Journal of Public Health* 95, 1568–1574.

Hastings, L., McDermott, L., Angus, K., Stead, M. & Thomson, S. (2006). The extent, nature, and effects of food promotion to children. World Health Organization. Accessed online at www.who.int/dietphysicalactivity/publications/Hastings_paper_marketing. September 10, 2009.

Hiroi, N. & Agatsuma, S. (2005). Genetic susceptibility to substance dependence. *Molecular Psychiatry* 10(4): 336–344.

Hitchings, E. & Moynihan, P. (1998). The relationship between television food advertisements recalled and actual foods consumed by children. *Journal of Human Nutrition & Dietetics* 11(6): 511–517.

Hoek, J. & Gendall, P. (2006). Advertising and obesity: A behavioral perspective. *Journal of Health Communication* 11(4): 409–423.

Hyde, M. J. (2008). *Perfection, postmodern culture, and the biotechnology debate.* New York: Pearson/Allyn and Bacon.

Hyde, M. J. & King, N. M. P. (2010). Communication ethics and bioethics: An interface. *Review of Communication* 10(2): 156–171.

Jukes, D. (n.d.) European Union (EU) Food Regulations and Standards Context, Implementation, and Cross-Border Implications. Available from: <http://www.farmfoundation.org/news/articlesfiles360-Djukes.pdf>

Keith, S. W., Redden, D. T., Katzmarzyk, P. T., Boggiano, M. M., Hanlon, E. C., Benca, R. M., et al. (2006). Putative contributors to the secular increase in obesity: Exploring the roads less traveled. *International Journal of Obesity* 30(11): 1585–1594.

Krcmar, M. & Farrar, K. (2009). Retaliatory aggression and the effects of point of view and blood in violent video games. *Mass Communication and Society* 12(1): 115–138.

Krcmar, M., Helme, D. & Giles, S. M. (2008). Understanding the process: How mediated and peer norms affect young womens' body esteem. *Communication Quarterly* 56(2): 111–130.

Krcmar, M. & Lachlan, K. (2009). Aggressive outcomes and videogame play: The role of length of play and the mechanisms at work. *Media Psychology* 12(3): 249–267.

Kunkel, D. & Gantz, W. (1992). Children's television advertising in the multichannel environment. *The Journal of Communication* 42(3): 134–152.

Larwin, K. H. & Larwin, D. A. (2008). Decreasing excessive media usage while increasing physical activity: A single-subject research study. *Behavior Modification* 32(6): 938–956.

Lawrence, R. G. (2004). Framing obesity: The evolution of news discourse on a public health issue. *Harvard International Journal of Press/Politics* 9(3): 56–75.

Lobstein, T., Baur, L. & Uauy, R. (2004). Obesity in children and young people: A crisis in public health. *Obesity Reviews* 5(s1): 4–85.

Lvovich, S. (2003). Advertising and obesity: The research evidence. *Young Consumers: Insight and Ideas for Responsible Marketers* 4(2): 35–40.

Marshall, S. J., Biddle, S. J. H., Gorely, T., Cameron, N. & Murdey, I. (2004). Relationships between media use, body fatness and physical activity in children and youth: A meta-analysis. *International Journal of Obesity* 28(10): 1238–1246.

Mokdad, A. H., Ford, E. S., Bowman, B. A., Dietz, W. H., Vinicor, F., Bales, V. S., et al. (2003). Prevalence of obesity, diabetes, and obesity-related health risk factors, 2001. *Journal of the American Medical Association* 289(1): 76–79.

Mokdad, A. H., Serdula, M. K., Dietz, W. H., Bowman, B. A., Marks, J. S. & Koplan, J. P. (2000). The continuing epidemic of obesity in the United States. *Journal of the American Medical Association* 284(13): 1650–1651.

Obama, B. The White House. Office of the Press Secretary (2010). Remarks by the president in State of the Union Address, Washington, D.C. Available from <http://www.whitehouse.gove/the-press-office>

Ogden, C. L., Carroll, M. D., Curtin, L. R., McDowell, M. A., Tabak, C. J. & Flegal, K. M. (2006). Prevalence of overweight and obesity in the United States, 1999–2004. *Journal of the American Medical Association* 295(13): 1549–1555.

Ogden, C. L., Flegal, K. M., Carroll, M. D. & Johnson, C. L. (2002). Prevalence and trends in overweight among US children and adolescents, 1999–2000. *Journal of the American Medical Association* 288(14): 1728–1732.

Powell, L. M., Szczypka, G. & Chaloupka, F. J. (2007). Exposure to food advertising on television among US children. *Archives of Pediatric & Adolescent Medicine* 161(6): 553–560.

Program for the Study of Media and Health (March, 2007). #7618. A Kaiser Family Foundation Report, Food for Thought: Television Food Advertising to Children in the U.S. Available from <http://www.kff.org/antmedia/7618.cfm>

Sallis, J., Owen, N. & Fisher, E. (2008). Ecological models of health behavior. In K. Glanz, B. K. Rimer, & K. Viswenath (Eds.), *Health behavior and health education: Theory, research, and practice* (4th ed., pp. 465-482). San Francisco, CA: Jossey-Bass.

Salwen, N. M. B. (1998). Perceptions of media influence and support for censorship: The third-person effect in the 1996 presidential election. *Communication Research* 25(3): 259–285.

Serdula, M. K., Ivery, D., Coates, R. J., Freedman, D. S., Williamson, D. F. & Byers, T. (1993). Do obese children become obese adults? A review of the literature. *Preventive Medicine* 22(2): 167–177.

Taras, H., Sallis, J., Patterson, T., Nader, P. & Nelson, J. (1989). Television's influence on children's diet and physical activity. *Developmental and Behavioral Pediatrics* 10(4): 176–180.

The Center for Consumer Freedom (n.d.). *An Epidemic of Obesity Myths*. Available from <http://www.obesitymyths.com>

Tsai, A. G. & Wadden, T. A. (2005). Systematic review: An evaluation of major commercial weight loss programs in the United States. *Annals of Internal Medicine* 142(1): 56–66.

Williams, J., Achterberg, C. & Sylvester, G. (1993). Targeting marketing of food products to ethnic minority youths. In C. Williams & S. Kimms (Eds.), *Prevention and treatment of childhood obesity: Annals of the New York Academy of Sciences*, vol. 699. New York: New York Academy of Sciences. 107–114

Young, B. (2003). Does food advertising make children obese? *Young Consumers: Insight and Ideas for Responsible Marketers* 4(3): 19–26.

Zoeller, R. F. (2009). Physical activity, sedentary behavior, and overweight/obesity in youth: Evidence from cross-sectional, longitudinal, and interventional studies. *American Journal of Lifestyle Medicine* 3(2): 110–114.

9 How We Feel with Metaphors for Genes

Implications for Understanding Humans and Forming Genetic Policies

Celeste M. Condit

PRECIS[1]

Critical analysis of discourse about genetics has, to date, focused primarily on its ideological components. This essay synthesizes data on public use of metaphors about genetics that shows emotive dimensions of discourse may also play a substantial role in determining people's understanding of, talk about, and response to genomics. The essay reports a study showing the ways in which members of the public explain their preference for particular metaphors based on how those metaphors make them feel. It also reports a study showing that lay preferences for metaphors differ from the expert metaphors circulated in the mass media. The essay offers an account of those differences based in the different emotional work that different metaphors do for experts and for members of the public, given their different situation in relationship to genetics knowledge. These findings imply the need for greater attention to the role of emotion in the uptake of genetic technologies such as personalized genomic testing and genetic engineering, and the need to form policy based on our experiences of these feelings rather than merely on putatively *a priori* principles.

INTRODUCTION

The rise of the "genetic revolution" has been accompanied by substantial scrutiny of public (Ceccarelli 2004; Lynch 2008; Nelkin and Lindee 1995) and expert (Fox Keller 2002; Journet 2005; Syed et al. 2008) discourses about genetics. As this essay will detail, most of this scrutiny has attended to the ideology or accuracy and sufficiency of these discourses. Arguably, however, academic understandings of the nature of humans and the bases of their interrelationships is undergoing a paradigm shift, from what has been called the "rational" or "Enlightenment" account to what might be called a biosymbolic model. This essay shows data highlighting the role of

emotions (a major component of the biosymbolic model) in human communication about genetics, specifically in metaphors, and it suggests the
importance of giving greater attention to emotion in research and policy
related to future genetic technologies.

The essay begins by summarizing the differences between the rational
paradigm and the biosymbolic paradigm, illustrating the ways in which
critique of metaphors about genetics has focused on rational models. It then
presents data showing that lay people account for their preferences among
metaphors about genetics based on how the metaphors make them feel,
rather than primarily based on the correctness of the analogical structure
offered by a metaphor. This finding generates a hypothesis that the preferred metaphors of experts and non-experts would differ, because experts
and non-experts are situated differently with regard to the preferred "coding" metaphors of experts. Two lines of data are reported that confirm that
supposition. The essay closes with a brief exploration of the implications
of the biosymbolic model of human beings for the generation of policy and
practice regarding genetic technologies.

HUMANS ARE NOT "THE RATIONAL ANIMAL" (THANKFULLY!)

In the past two or three decades, scholars have pointed out the ways in
which the "rational world paradigm" (Fisher 1987) has been a mainstay
of the Western academy (Perelman & Olbrechts-Tyteca 1969; Venn 2006).
They have noted that, at least since Aristotle's definition of humans as the
"rational animal," academic luminaries ranging from Descartes to Kant
to Habermas have replayed the assumption that human decision-making
and knowledge-production were driven primarily by rational capacities
that other animals lacked (Little 1995). Within this Lockean (1970) model,
language functions as a conveyer belt to bring "ideas" from the (rational)
mind of one person to the mind of others. The most important thing about
language, on this model, is its accuracy in representing the components of
the world that people are thinking about.

The rational worldview about humans and their communication practices has not gone unchallenged in the history of Western thought. Rhetorical theorists ranging from Gorgias to George Campbell (1988) to Kenneth
Burke (1966) have offered counter-definitions, which have emphasized
the linguistic capacities of humans, rather than their rational capacities.
Instead of focusing on the referential character of language, these theories
have emphasized the capacity of language to move people to action and
arouse passions such as delight. More recent versions of the post-rational
approach have begun to see language as fundamentally about maintaining
relationships rather than solely about discovering truths. These perspectives, based in rhetorical theory, emphasize that an ideal of human perfection that takes rationality as its central principle would reduce human

beings to computational machines (Weaver 1965). Instead, they encourage ideals of humanity that are multidimensional, including emotional richness and intellectual, aesthetic, and social creativity.

A separate challenge to the rational view of humans has come from biology. Charles Darwin (1879/2004) himself disputed the idea that humans had rational capacities that were different in kind from those of other animals. The past decade has seen the substantiation of Darwin's claims, as a wealth of data has now shown that other animals have almost every imaginable form of rational capacity: primates have been shown to use transitive logics (Cheney & Seyfarth 2007), chimpanzees are as good at or better than humans at numerical/spatial memory (Inoue & Matsuzawa 2007), rooks can analyze problems and generate two-part tools to solve them (Bird & Emery 2009), and dogs, parrots, and chimpanzees can generate vocabularies with hundreds of words and assemble them into basic grammatical relationships (Kaminski, Call & Fisher 2004; Pepperberg 1998). Even the claim that only humans have "consciousness" appears to be crumbling (Griffin & Speck 2004).

Today there is, at the least, good reason to consider the idea that the "rational animal" is not as good a definition of human beings as the one offered by Kenneth Burke (1966), who suggested humans should be defined as "symbolizing animals." From this view, humans are beings whose actions are determined by both biological and supra-linguistic qualities, and whose immense capacities for tool-making, reasoning, and aesthetic creativity are primarily products of the superlative levels of complexity in human communication and other cognitive capacities, but not exclusively governed by those capacities. This view replaces the idea that language is fundamentally to be understood and judged as a tool of reference (i.e. how well it clearly represents the things of the world) with the idea that language is substantially a means of making things happen in the world and maintaining human social relations (Cheney & Seyfarth 2007; Little 1995).

Burke provided groundwork for the merger of a biological and symbolic view when he argued that language should not be examined primarily in terms of its referential sufficiency, but instead in terms of its motivational characteristics. Burke himself, however, focused almost exclusively on elaborating human symbolic systems, and so failed to come to terms with exactly how the "animal" or "bio" part of the biosymbolic approach to humans might shape linguistic processes. That gap is rapidly being filled in by what has come to be called "the affective turn" (Clough & Halley 2007). Research in psychology, media poetics, neuroeconomics, marketing, political science, and elsewhere is beginning to develop the case that human decisions are not driven by rational calculus or unalloyed self-interest, but rather by complex feelings, emotions, affects, cognitive biases, perceptual biases, and linguistic cues, among other components (Alford & Hibbing 2004; Camerer, Loewenstein & Prelec 2005; Hauser, 2006). It will be a massive task to sort out all the dimensions of this profusion of research agendas. However, the accumulated evidence has grown beyond the point at which its challenge to the rational world paradigm can simply be ignored if one's goal is to try to understand human beings and to build theories and policies appropriate to human social action. I therefore propose in this essay to bring this perspective to bear on the analysis of genetic discourse and its implications for public policy. Because metaphors are, in a sense, the wordiest of words, they form an appropriate starting point.

IDEO*LOGICAL* DOMINANCE IN ANALYSIS OF GENETIC METAPHORS

I have elsewhere (Condit 2009) provided an overview of the many essays written about genetic metaphors. These essays tend to concur that the most common metaphors for genetics are coding metaphors, including maps, books, languages, and especially blueprints, and that they are undesirable (e.g. Hedgecoe 1999; Fox Keller 2002; Lippman 1992). I wish here to suggest that these analyses overwhelmingly arise from a version of the rational world paradigm. They assume that metaphors should represent the world correctly, and they argue that coding metaphors do not do so. Their authors might well respond that they are offering ideological analyses rather than rational world analyses, and they would point to the views of Marx and Engels, rather than Descartes, Kant, and Locke as progenitors. Marx and Engels argued that ideas arise from economic structures rather than universal truths. However, the contrast to emotional analyses reveals that even ideo*logical* analyses grounded in Marx and Engels' work are substantially influenced by the rational world paradigm. The dominant mode of such critiques is the demonstration of logical contradictions and false universalization (Giddens 1979). Ideological critiques seek to delegitimize discourse produced by dominant elites by showing that such discourses are not universal and that they are logically contradictory. Such critiques maintain referentiality as the taken-for-granted standard of language use. These tendencies are evident in the major critiques of the blueprint metaphor for genetics.

Abby Lippman has provided many perceptive critiques of genetics and its delivery. In addition, she provided one of the earliest criticisms of the blueprint metaphor when she wrote that "In addition to the reductionism inherent in the blueprint metaphor is a problematic notion of determinism" (Lippman 1992, p. 1471). This is an ideological analysis dependent on the notion that the goal of language is accurate representation of an external world. The basis of her argument for the inadequacy of the blueprint metaphor is that it is reductionistic—it does not capture the full range of variables involved in producing human outcomes. Moreover, because of this, it invites what she identifies as determinism (by which she arguably means biological determinism), which she believes to be an *incorrect* characterization of human nature.

This analysis was extended by Ruth Hubbard and Elijah Wald in their trenchant and extensive critique of genetics. They wrote that this metaphor "is testimony to the hierarchical models they [geneticists and their supporters] use rather than a description of the ways in which organisms function" (Hubbard and Wald 1993, p. 64). This critique faults the blueprint metaphor on the grounds that it is not a description. Applying the now-standard approach of ideological criticism, they argue that instead, the metaphor is grounded in the interested discourses of geneticists—for whom hierarchy is a favorable worldview. To be fair to their work, they do not stop merely at decrying the failed referentiality of the metaphor. They also add an effects analysis, which explores what the metaphor might do in the world, and such analyses can be part of a post-rationalist paradigm. However, even that analysis is grounded in the meanings that Hubbard and Wald believe lie in the words, and this approach falls back on the assumption that there are stable, univocal meanings associated with words—a key tenet of the rational world paradigm.

To point to one final example, one of the most emotionally rich analyses of the genetic revolution that has been written is that authored by Barbara Katz Rothman. It is surprising, therefore, that her analysis of the blueprint metaphor fits squarely within the rational world paradigm. She criticizes the blueprint metaphor for its referential insufficiency when she complains that the metaphor " . . . doesn't have time built into it; blueprint isn't about growth, and so it isn't a good analogy for DNA" (Katz Rothman 1998, p. 23).

These examples all come from individuals whose writing otherwise indicates that they would likely not be supporters of the rational world paradigm. This indicates how deep that paradigm runs in our consciousness. Indeed, I fully shared those approaches and perspectives until a research project caused me to be able to listen to human symbolizing in a way that took seriously the emotive components of human language.

STUDY 1: METAPHORS ARE CHOSEN FOR THE FEELINGS THEY GENERATE

In the winter of 2008, our research team conducted interviews with 13 people, all of whom had an annual income equal to or less than $35,000 and no more than one year of college education. Details of the interview protocol are reported in Condit (2009) and some of the data reported here is also reported there. This practice of multiple publication of pieces of data is sometimes derided as "shingling" in the rational world view. The rational world paradigm presumes that publication is universal, out of time and audience independent; once published a piece of data is accessible to everyone, everywhere, and the time necessary for access doesn't matter, and mode of presentation does not need to be adapted to different audiences. The biosymbolic perspective suggests instead that messages are delivered to specifically configured audiences in specifically tailored ways. Thus, the mere publication of bits of data is less important than the way multiple pieces of data are configured within messages at particular places and addressed to particular audiences. As an advocate of the biosymbolic model, I am proceeding (with the editors' permission) according to the assumption that reassembling (parts of) data in different publication venues is a valuable and appropriate practice.

Our interviewees listened to two messages about different causes of health. After each message we asked them to choose one of four metaphors for describing the relationship of genes and behavior (described below) that had been selected from previous interviews. After participants made their selection, we asked them to explain their choice.

The metaphor that our research team thought best represented an "interactive" understanding of the relationship between genes and health behaviors was the metaphor of "snowball," which we selected from a participants' use of it in a prior interview. We presented it in the following statement, "Word #2 is 'Snowball': Unhealthy behaviors make the effects of unhealthy genes SNOWBALL to produce disease." We selected the metaphor based on its conceptual qualities—the snowball seemed to offer the most appropriate analogy to genes/behavior: just as the amount and type of snow and the steepness of the slope of a hill interacted to produce the speed and size of a snowball rolling down a hill, genes and behavior interacted to produce health outcomes, through time.

The second metaphor—"trigger"—we described with the sentence, "Unhealthy behaviors can TRIGGER a gene for a disease." This metaphor is occasionally used by both specialists and non-specialists. Its analogical content has the virtue of including both genes and behaviors as crucial components of the outcome, but it has the undesirable feature of implying that the interaction is a one-time phenomenon, rather than developing over time.

The third option was the "additive" metaphor, which we had found previously to be highly common (Condit et al., 2009). It is a non-interactive conceptualization of genes and behaviors as independent contributors to health outcomes. We described it in this way, "Word #3 is 'Add on': The risks of unhealthy behaviors are ADDED ON top of the risks from unhealthy genes." The final choice was the genetically deterministic one, "Word #4 is 'Outweigh': Unhealthy genes OUTWEIGH behaviors."

To our complete surprise, the participants' explanations of their preferences of metaphors did not reference analogical content. Rather than discussing the degree to which gene/behavior were like or unlike snowballs, triggers, adding up, or outweighing, participants talked about the different feelings that the metaphors generated in them as the reason for their preferences.

The most explicit and extended account of this basis for choice occurred in the following interchange (Participant 31):

I: I think a trigger ___ you automatically think of a gun. So, I would say trigger is a negative word. Snowball. When you think of snowball you always think of the, the avalanche, especially when you use it in that context, so that would be covering of the whole thing. Um, not feeling snowball. I actually like add on. Add on is not too frightening but it is enough to be concerned with, and I forget the last one.

M: Outweigh

I: Outweigh, hmmm I'm not feeling the word outweigh.

M: Not outweigh?

I: No, just not feeling it, but add on.

Other participants gave shorter replies, and they favored different metaphors, but their answers were likewise based on how the metaphor made them feel rather than on the cognitive content of the metaphors being offered (that is, whether or not a concept-based analogy between genes/behavior and snow/hill could be built). Person 10 said she chose "add on" because "it was not as scary as those other words." She elaborated, "So, if you would just say add on, people wouldn't run from it, they wouldn't shut down, you know? Just add on. Whatever you're doing, just add this on, trigger and snowball are overwhelming . . ." The fear generated by the snowball metaphor is palpable in another response:

> And with the snowball, if you continue to keep doing something, snowball what it does is, when it goes down hill it keeps increasing, and it keeps increasing and it keeps, you know, until it becomes this big old ball that's out of control and it's just rolling and hurting everybody and rolling, you know, just _____ stuff, and (pause), um . . .

A final example came from someone who explicitly endorsed the "add on" conceptual structure ("on top of"), while nonetheless adopting the snowball metaphor for its emotional value. This participant said, "Because it's, um, your diet on top of your genes and, accelerates it. Snowball real fast." (P56)

In so far as they were able to articulate the bases for their choices, our participants selected their descriptions of the relationship of genes and behavior based on how specific metaphors made them feel, not primarily on the conceptual or logical structure of the relationships between the elements in the metaphor. Conceptual or logical structures were still inevitably relevant, as semantic contents that are *prima facie* inappropriate (e.g. "genes are like spaceships") would likely be rejected as nonsensical. However, prior research has indicated that affects may actively drive the convergence process among beliefs, attitudes, values, and decisions (Finucane et al. 2000), and in this case, the affective components of the metaphors appeared to serve that role.

STUDIES 2 AND 3: COMPARING LAY AND EXPERT METAPHORS

Having been startled out of my complacent assumptions about the bases on which people selected metaphors, my interest was intensified in the question of whether experts and non-experts could really be expected to use the same metaphors for genetics. To explore that question further, I turned to data already at hand, but not addressed with regard to this question.

The first question one needed to settle in comparing expert to lay metaphor preference regarded the representativeness of the results of the 13 interviews we had conducted. We had previously asked 1,218 people via a computer-based survey conducted by Knowledge Networks which of the four metaphors they thought best represented the relationship between genes and behavior (details of the study are available in Condit and Shen (2011). Approximately 18% chose "trigger," 17% chose "snowball," 49% chose "add on," and 11% chose outweigh (the term used there was "dominate"). This indicated that the majority of people did not have an "interactive" perspective on the relationship between genes and behavior (although a third did have such a view). More importantly for present purposes, the critics identified above had claimed that the "blueprint" metaphor has the same analogical structure as the "outweigh" metaphor. So this result provided one line of evidence that lay metaphoric preferences were different from those of experts in terms of their cognitive assumptions about the relationships of genes and behaviors.

This data did not directly ask people about the "blueprint" metaphor. Given the now-obvious (to me) importance of emotional resonances, one could no longer assume that the analogical structure was all that mattered. Moreover, research had previously indicated that lay people themselves interpret the blueprint metaphor in different ways (Condit et al. 2002). So, though that data gave us one measure that suggested lay people did not prefer the same genetics metaphors as experts did, we decided to turn to a corpus of lay discourse that we had previously gathered to address the question of lay people's choice of metaphors in a different fashion.

In the spring and fall of 2006 our research team had conducted 96 interviews with lay people. Details of the study are reported in Condit et al. (2009). The study did not prompt people with particular metaphors or ask about metaphors directly. Instead, participants had simply been asked to talk about particular people they knew who had heart disease, depression, diabetes, or lung cancer. After they had expressed their "top of the head" views, they were prompted to think about genes or behavior and their relationship as causal factors. This provided a fairly large corpus of talk about genetics in which the metaphors were "respondent generated"—that is, respondents were not asked to respond to specific metaphors, or cued with them, but instead employed (or more commonly did not employ) metaphors as a matter of their own language choices. The fact that respondents' use of metaphors was not driven by a research agenda to articulate particular metaphors is substantiated by the relative paucity of "live" metaphors that were countable by our research team (only 67 different uses of "live" metaphors by 38 of our 96 participants).

The metaphors that participants chose were surprising. The most common cluster of metaphors were those describing genes as a virus, infection, or disease (n=22). These included direct comparisons of genes to HIV/AIDS as well as direct statements that genes were themselves the disease. The second most common set of metaphors were "trigger" metaphors, which used terms such as trigger or activate (n=15). Three groups of metaphors all occurred six times each. These included metaphors that described behaviors as "fuel" added to the "fire" that is the gene (n=6), various game or gambling metaphors (n=6; e.g. playing Russian roulette or having a "strike" against you), and treatment of the gene as "blueprint" or "map" or "code" (n=6).

These data provide the best available evidence about the metaphors people actually might be using for talking about and understanding genes, and at least in these contexts, coding metaphors were not predominant for members of the general public.

ACCOUNTING FOR EXPERT AND PUBLIC FEELINGS ABOUT METAPHORS

Given that emotions appear to be crucial to choice of metaphor, an appropriate next step is to try to invent a mode of analysis that accounts for metaphors not in terms of what they mean, but instead in terms of the kinds of emotions they may generate. There will likely be instant objections to such a mode of analysis, on the grounds that we can't "really know" what people are feeling. In this regard, however, affect-based analysis is no different from the now well-entrenched ideological analysis. Multiple audience-based studies have shown that what critics think texts mean is not always the same as what audiences think texts mean (Morley 1980; Radway 1984). Ideological criticism has largely been able to hide that discrepancy because ideological critics tend to write for like-minded academics, for whom particular words and metaphors are most likely to share the meanings advanced, but this sleight of hand is hardly in its favor. The problems of interpretive evidence are therefore no different for affective and ideological criticism; they are just more obvious and less veiled in the case of affective analysis. One might argue that this is a virtue of an affective analysis—it forces us to be more honest about the evidentiary force of our claims.

Neither affective nor ideological criticism can pass the evidentiary tests of the hypo-deductive quantitative method. Their intellectual function is not to test, but to advance possibilities and create syntheses. The standards of judgment employed should not be certainty, or even extremely high probability, but merely plausibility based on coherence and preliminary evidence, as well as potential impact. When a reading becomes relevant and controversial, quantitative methods can sometimes be brought to bear if validation is desirable. However, like ideological criticism, the focus of an affective analysis is not simply on individual response—how a particular individual or set of individuals might respond to a metaphor. Because critical analysis operates at the level of social significance and social explanation rather than staying at the individual psychological level, an affective critique is interested in describing what Raymond Williams (1977/1985) has labeled the "structures of feeling." An affective critic thus will attend to issues of social positioning as these open up and close down the likelihood of particular kinds of emotive responses. With those criteria in mind, I offer a set of tentative affective analyses that provide potential explanations not only for why it might be that experts and the public differ in their affection for coding metaphors, but also for why a metaphor as surprising, peculiar, and even off-putting as the "virus" might be an affectively supported choice among some members of the general public in some contexts.

Genetics Professionals

For geneticists, the potential for positive feelings to be generated by coding metaphors may rest on the promise they imply to be able to *read* human beings. Such a stance offers a feeling that one is above other humans, understanding and predicting their actions. On a positive reading, this might be associated with the possibility of helping to heal and to cure. On a more sinister reading, this may seem to offer power over others. Although one might rush to deplore such a feeling, there is substantial evidence that such status drives are common and have powerful influences upon human beings (Cheney & Seyforth 2007). Another dimension of appeal of the reading metaphor may lie in its association with "ease," as reading seems like a physically un-demanding and rapid activity for well-educated professionals. The strength of the emotional attraction of the promise of easy power to know the secrets of other humans' bodies may explain why there has been such sustained excitement over genetics research, even in the absence of much in the way of major applications. Marxist-style critiques would claim that the excitement derives solely from the large amount that geneticists get paid for trying to read these codes, and that is surely part of the story, but these scientists would get paid handsomely for any beneficial technology they could promise. The ability of genetics to *sustain* attractiveness even in the *absence* of a large body of direct and visible applications has been, I would suggest, due to the emotional attractiveness of the promise of a power that seems so potent and broad-reaching—to read the human being. Hubbard and Wald (1993) were thus correct to identify "hierarchy" as at work in the experts' choice of coding metaphors, but it is not necessarily because these experts subscribe to a "hierarchical" world-view, but rather because the metaphor generates for experts the positive feelings of being on the top of a hierarchical relationship structure.

Media Professionals

When critics argue that coding metaphors dominate representations of genetics, they often conflate the contribution of geneticists with those of reporters. Indeed, it is not always clear whether the use of a particular metaphor appearing in public media was stimulated by a reporter or by a geneticist. In all cases, however, for a metaphor to be retained in an article by the reporter who writes it, there must be some resonance with that metaphor for the reporter. An account of the emotional affiliation between reporters and coding metaphors seems ready to hand. The life business of journalists is the transmission of codes. Coding metaphors are thus familiar and comfortable. Centralizing coding metaphors also centralizes what journalists do. It seems reasonable to suggest that the prominent role of coding metaphors for genetics in the mass media is therefore a product of the overlap between the affective resonance of these metaphors for journalists and geneticists.

General Public

If the most commonly circulated metaphors for genetics are coding metaphors, why then don't members of the public adopt this vocabulary? If genetics—reading the human code—were not perceived as something enormously complicated that only specialists will be able to do—then people in general might feel about "genetic blueprints" the same way that geneticists do. But genetics is widely understood as a technology that will always only be under the direct control of specialists. It will be an expert-controlled technology, rather than a user-controlled technology. Non-physicists can get excited about a DVD player, because they can operate it themselves. They don't need to understand the underlying physics to use the machine. So the DVD is something whose feeling structure appeals to the technological generalists who will use it in at least some of the same ways that it appeals to the specialists who create it. Both can get excited over the powers and opportunities DVD technologies may provide.

In contrast, non-specialists do not know—and cannot reasonably expect to know—how to read genetic "blueprints." People in general don't expect to be able to "read" the map of the genome, or the book of life, themselves. Reading scientific facts is not experienced as easy by the majority of the U.S. general public, but rather as difficult, demanding, even impossible. As a result, the feeling structure created by "genetic blueprints" is one in which you are asked to imagine someone having understanding of the blueprint of YOU that you can't share. The threatening character of such a feeling may well explain why the structure of feeling invited by the blueprint metaphor and most of the coding metaphors has less appeal to the general public than to experts. Indeed, the undesirability of giving experts such control also leads some of our interview participants to reject actively the metaphoric clusters' appropriateness, denying that experts can know these things: Participant 1555 said:

> Nobody knows. Right now we're talking about a human being like a car, like a gene is all the information, but I know exactly, no scientific, nobody, have the complete answer about anything. Everybody figure out, they make comparation, they compare this, and compare the other thing. They have some idea about the gene, they talking about, but nobody have the complete knowledge of that . . . the scientists they say something today, and in one week they're going to say something else.

There are therefore strong feelings that explain why most members of the general public do not spontaneously employ the coding metaphors for genetics. It does not feel good to be the bug under someone else's microscope. But what possible positive feeling could a metaphor like "virus" offer? An answer can be constructed from a comparison to the feelings likely to be associated with the other metaphors people seem to have available to them.

The other major metaphoric clusters include gambling (e.g. "it's kinda like playing Russian roulette. They are taking great big risks" (P102)), fuel on the fire (e.g. "It's like adding fuel to the fire. It's like pouring gasoline on a fire" [P95]), and trigger ("that's a risk that you find in your body. It's just—be careful not to trigger it" [P1092]). All of these rather different metaphors share at least one feature—they place blame on the individual for causing their own illness. Accepting that one is causing one's own serious illness is not likely to be a positive feeling. If people could reasonably expect to change their behaviors, then perhaps such an indictment might be tenable. But many studies of health communication campaigns have demonstrated that merely knowing that you are at risk does not change behavior very much for many people. Indeed, even people who have actually suffered heart attacks and therefore are at very high risk for recurrence comply with medical prescriptions typically at less than a 50% rate (Evangelista & Dracup 2000). This occurs because beliefs are motivated in the moment and belief selection is caused by feelings. Long-term scripts based primarily on abstract beliefs (e.g., "one should exercise") are not therefore primary or sufficient causal sources of behaviors (Condit et al. 2009; Camerer et al. 2005).

It seemed to me from listening to our interviews that many people feel that they cannot change their routine behaviors, given the total contexts of their lives. For people who feel this, the virus/disease metaphors may feel better than any of the other available choices. Viruses are *in* us, and they *may* manifest themselves sometime in the future. I may have bad cells, be diseased, and someday I'll feel sick from them, but that is the nature of life. After all, some day we really are all going to die. Moreover, it may be a long time before the viruses cause any problems, so I can temper the threat by pushing it off into the future.

Consequently, although the virus/disease metaphoric cluster may not constitute an ideal set of sentiments, for many people it may be the best available option. Normalizing genetic disease to viruses and the standard course that life takes may produce statements that sound fatalistic, but it avoids either taking blame for illnesses that you really may not be able to control or giving others scopic power over your life. Having been surprised multiple times already by interviews with the public, I am not advancing these speculations as definitive, but only as lines for exploration. I would suggest, however, that they provide a plausible account that deserves further investigation.

CONSOLIDATING EMOTION IN THE BIOSYMBOLIC PERSPECTIVE

This essay has provided additional grounds for supplanting a definition of human beings as the rational animal with an account of humans as biosymbolizers. This does not mean that humans are irrational. Most kinds of rationality are common to all complex biological creatures, and some are common to all biological creatures. It may even be the case that humans are (sometimes) capable of extending these basic rational capabilities, by using symbolic tools and due to sheer neural mass. However, the biosymbolic view insists that rational capacities are not separate from and above humans' other biological features, including what has gone under the heading of "emotions." As much psychological and neuro-imaging research has now established—human emotions are the substantive foundations of our reasoning and decision-making (Damasio 2006; Finucane et al. 2000). Emotion is not something to be stashed away in a closet or controlled by the rational capacities of a (superior) superego, id or noble charioteer. Instead, emotions are the basic ground of what enables us to prefer and to act. Consequently, we need to begin to grapple with how policies might be generated in ways that take account of, and respect, our emotional capacities, rather than in obeisance to putatively universal "rationally" grounded principles that do not sufficiently capture human beings (Hauser 2006; Thaler & Sunstein 2008).

Research on the biosymbolic nature of human beings has passed its infancy, but remains in the stormy adolescent period, where settled frameworks and coherent linkages have not been built. But we have enough shared experience to formulate a good-enough-for-now approach, which posits an alternative to the idea that with enough reason we can quash our emotions. Instead, we can imagine the perfection (or, rather, improvement) of humans not as computing machines, but rather as empathic, creative seekers. Such a view sets the trajectory of human improvement not as gathering more knowledge and restraining our feelings, but rather as learning to explore and expand our emotions in ways that enhance the shared structure of feelings. Such a view posits that our biologically induced selfishness cannot be best countered by teaching individuals to tamp down their emotions, but rather by teaching them to expand their (biologically enabled) empathic capacities through the imaginative abilities with which we are endowed through language and other symbolic systems. For example, instead of trying to counter our myopic focus on the present with principle-like injunctions to factor the long-term into our calculations, we might bring the future, vividly imagined and felt, into the present. This alternative trajectory also has implications specifically for genetics policy and research.

RESEARCH AND POLICY IMPLICATIONS

On the research front, I hope I have illustrated—in a tip-of-the-iceberg fashion—the possibilities for understanding how people will be moved by and motivated to use genetics information. A large literature on patients' understanding of and decision-making about genetics exists (Sivell et al. 2008), and some of this points to the "irrational" components of these processes as a failure in lay processing of genetics information. Such research thus starts with the framework that people should be rational information processors who use genetics information in a uniform and calculatively correct fashion, and it finds that this is not what people generally do. It is time to change the frame of reference to one that says: people are biosymbolic beings. They will utilize genetic information in accord with short-term biases, social comparison processes, and other species-typical processing characteristics, and they will use the information in ways that make them feel better about their experiences and perhaps their possibilities. Now what does that mean for what genetics might do, for how it will be used, and for the kind of support and materials that people ought to be provided?

This is to say, more broadly, that all policies about genetics should, at the very least, systematically and rigorously factor emotion into their considerations. Most policy formulations have been debated on the basis of what are presumed to be universalizable principles. But if humans are not creatures that operate on the basis of a coherent and stable hierarchy of principles, then policy formulated on such bases are unlikely to fit the realities of human lives. There has been some tentative efforts at using emotion as a basis for policy formation (Junker-Kenney 2005; Kass 1997), but in these cases, the emotions are imputed (parents will be like this; people respond to genetics like that) rather than empirically engaged. Once we admit that policy should be based on how human beings feel toward and use a technology instead of based on abstract, universal principles, we can no longer simply impute the emotions to others at a distance (in part because we tend to produce overly simple projections). We have to engage in the careful, empirical, systematic work needed to explore the range of emotions as they exist, and perhaps their causation.

An example of such work has been provided for us by Marli Huijer's (2008) exploration of the ways in which families who had experienced inherited cancer syndromes (especially breast cancer) talked about their decisions to undergo, or not undergo, prenatal genetic diagnosis for the alleles associated with these cancer risks. She found that some parents rejected prenatal selection because the web of experiences led them to believe that "I was strong enough to handle breast cancer; she [my daughter] will be too" (p. 4). Other sets of parents chose PGD to foreclose BRCA-based susceptibility to breast cancer for their children because they felt that the suffering was too great: "I would love to have children, but this disease should not be transmitted. It must stop at us" (p. 3). Huijer emphasizes that "each story is unique" (p. 6). The different decisions made by different families came out of complex relational webs that contained many pertinent factors, including the severity and age-of-onset of disease experienced by family members, but also social support, world-view, and values. As Huijer reveals, these complex webs exceed the simplistic processes involved in articulating a single principle to guide decisions. Instead, these complexes came to be embodied in persons through their life experiences, and the information gained through these experiences was weighed and expressed through emotions.

Huijer's analysis should be contrasted to that of Jürgen Habermas, who has accepted the mantle of the late Enlightenment perspective (2003, p. 26). Habermas argued for a principle that would distinguish genetic enhancement from genetic therapy based on a careful philosophic analysis of principles associated with what he identified as human nature. To sustain this argument, he projected a set of emotions in which all parents contemplating genetic therapy had emotional sets focused on the well-being of their children and all parents contemplating genetic enhancement had emotional sets focused on the well-being of themselves (at the expense of or in neglect of their children). His use of principle to specify policy was thus based in a particular set of assumptions about people's emotions, but the set of emotions that he posits is shown, by Huijer's empirical work, to be false. There is no easy alignment of parental emotions with the healing/enhancement distinction. If there is no consensus about genetic engineering for highly elevated risks for breast cancer and ovarian cancer, it seems unlikely that one can treat "healing" as Habermas does, as a category of universal consent.

If universal principles have proven to be delusions, how then should we formulate policy about genetic selection? I would suggest that we do so on the basis of gradually accumulated experience. Rather than laying out fixed principles in advance (i.e. "permit treatment, forbid enhancement"), let us see how families experience the use of the currently limited genetic technologies for selection. Prenatal genetic selection seems a fortuitous place for our species to begin to gather experience, because it enables families to make choices only with regard to conditions with which they have already had some experience. Because they cannot insert new genes, but only decide whether or not their children should live with particular genes they have themselves experienced, parents have emotionally saturated experiences to bring to bear in making such selections. We should attend carefully to how such processes play out. What kind of structures of feeling actually come into play between parents and children created in these ways?

I suggest that none of us can know the answer to this question in advance. Once our societies have more experience with these limited, relatively well-informed selections, we will then be ready to advance judgments about whether broader ranges of choices (which we now call "enhancement") are likely to produce for us structures of feelings among humans that we would like to permit or support.

The dream of the Enlightenment was that humans could produce a coherent calculation of the principles necessary to explain and control the universe. This dream has foundered in the face of human nature, because human symbolic capacities have enabled the species to create novel forms at a dizzying rate. We can create new forms far faster than we can generate principles that encompass them as explanations, and the complexity of what we create defies any single rule. As the species faces the possibility of also modifying the biological inputs to its nature, the assumptions, dicta and hopes of the Enlightenment—based as they are in assumptions of stasis—necessarily prove insufficient. New vectors of approach, formed in light of growing sensitivity to a broader range of human characteristics, and tolerant of incremental adaptation, may be worth consideration.

NOTE

This essay is dedicated to Barbara Bowles Biesecker and Barbara A. Biesecker, as a testimony to the power of relational networks. I would never have been able to travel the paths I have traveled in the past decade and a half without their willingness to connect people together. Many thanks. Some of the material in this chapter appeared in the journal *Genomics, Society and Policy* (Condit 2009). I am grateful to the editors for their permission to reproduce some parts of that article.

REFERENCES

Alford, J. R. & Hibbing, J. R. (2004). The origin of politics: An evolutionary theory of political behavior. *Perspectives on Politics* 2(4): 707–723.

Bird, C. D. & Emery, N. J. (2009). Rooks use stones to raise the water level to reach a floating worm. *Current Biology* 19(16): 1410–1414.

Burke, K. (1966). *Language as symbolic action*. Berkeley, CA: University of California Press.

Camerer, C., Loewenstein, G. & Prelec, D. (2005). Neuroeconomics: How neuroscience can inform economics. *Journal of Economic Literature* 43(1): 9–64.

Campbell, G. & Bitzer, L. F. (Eds.) (1988). *The philosophy of rhetoric.* Carbondale, IL: Southern Illinois University Press.

Ceccarelli, L. (2004). Neither confusing cacophony nor culinary complements: A case study of mixed metaphors for genomic science. *Written Communication* 21(1): 92–105.

Cheney, D. L. & Seyfarth, R. M. (2007). *Baboon metaphysics: The evolution of a social mind.* Chicago: University of Chicago Press.

Clough, P. T. & Halley, J. O. (2007). *The affective turn: Theorizing the social.* Durham, NC: Duke University Press.

Condit, C. M. (2009). Dynamic feelings about metaphors for genes: Implications for research and genetic policy. Special issue: Genomics and metaphoric plurality. *Genomics, Society and Policy* 5:44-58.

Condit C. M. & Shen, L. J. (2011). Public understanding of risks from gene-environment interaction in common diseases: Implications for public communications. *Public Health Genomics*, vol. 14: 115–124.

Condit, C. M., Bates, B. R., Galloway, R., Givens, S. B., Haynie, C. K., Jordan, J. W., et al. (2002). Recipes of blueprints for our genes? How contexts selectively activate the multiple meanings of metaphors. *Quarterly Journal of Speech* 88(3): 303–325.

Condit, C. M., Gronnvoll, M., Landau, J., Shen, L., Wright, L. & Harris, T. M. (2009). Believing in both genetic determinism and behavioral action: A materialist framework and implications. *Public Understanding of Science* 2009, vol. 18 (6): 730–746

Damasio, A. R. (2006). *Descartes' error: Emotion, reason and the human brain.* London: Vintage.

Darwin, C. (1879/2004). *The descent of man, and selection in relation to sex* (2nd ed.). London: Penguin.

Evangelista, L. S. & Dracup, K. (2000). A closer look at compliance research in heart failure patients in the last decade. *Progress in Cardiovascular Nursing* 15(3): 97–104.

Finucane, M. L., Alhakami, A., Slovic, P. & Johnson, S. M. (2000). The affect heuristic in judgments of risks and benefits. *Journal of Behavioral Decision Making* 13(1): 1–17.

Fisher, W. R. (1987). *Human communication as narration: Toward a philosophy of reason, value, and action.* Columbia, SC: University of South Carolina Press.

Fox Keller, E. (2002). *Making sense of life: Explaining biological development with models, metaphors, and machines.* Cambridge, MA: Harvard University Press.

Giddens, A. (1979). *Central problems in social theory: Action, structure, and contradiction in social analysis.* Berkeley, CA: University of California Press.

Griffin, D. R. & Speck, G. B. (2004). New evidence of animal consciousness. *Animal Cognition* 7(1): 5–18.

Habermas, J. (2003). *The future of human nature.* Cambridge: Blackwell Publishing.

Hauser, M. D. (2006). *Moral minds: How nature designed our universal sense of right and wrong.* New York: Ecco.

Hedgecoe, A. M. (1999). Transforming genes: Metaphors of information and language in modern genetics. *Science as Culture* 8(2): 209–229.

Hubbard, R. & Wald, E. (1993). *Exploding the gene myth.* Boston: Beacon Press.

Huijer, M. (October 9–11, 2008). *Religiously inspired values versus women's lived experiences: The Dutch debate on embryo selection for hereditary breast cancer.* Paper presented at Towards a "Lingua Democratica" for the Public Debate on Genomics, Utrecht, International Expert Seminar of the University for Humanistics.

Inoue, S. & Matsuzawa, T. (2007). Working memory of numerals in chimpanzees. *Current Biology* 17(23): R1004–R1005.

Journet, D. (2005). Metapahor, ambiguity, and motive in evolutionary biology: W. D. Hamilton and the "gene's point of view." *Written Communication* 22(4): 379–420.

Junker-Kenny, M. (2005). Genetic enhancement as care or as domination? The ethics of asymmetrical relationships in the upbringing of children. *Journal of Philosophy of Education* 39(1): 1–17.

Kaminski, J., Call, J. & Fischer, J. (2004). Word learning in a domestic dog: Evidence for "fast mapping." *Science* 304(5677): 1682–1683.

Kass, L. R. (1997). The wisdom of repugnance. *New Republic* 216(22). Accessed online at <http://www.catholiceducation.org/articles/medical_ethics/me0006.html> Accessed July 19, 2009.

Katz Rothman, B. (1998). *Genetic maps and human imaginations: The limits of science in understanding who we are.* New York: W. W. Norton & Co.

Lippman, A. (1992). Led (astray) by genetic maps: The cartography of the human genome and health care. *Social Science & Medicine* 35(12): 1469–1476.

Little, M. A. (1995). Seeing and caring: The role of affect in feminist moral epistemology. *Hypatia* 10(3): 117–137.

Locke, J. (1970). *An essay concerning human understanding, 1690.* Menston, Eng.: Scolar Press.

Lynch, J. (2008). Geography, genealogy and genetics: Dialectical substance in newspaper coverage of research on race and genetics. *Western Journal of Communication* 72(3): 259–279.

Morley, D. (1980). *The nationwide audience: Structure and decoding.* London: British Film Institute.

Nelkin, D. & Lindee, S. (1995). *The DNA mystique: The gene as cultural icon.* New York: W. H. Freeman.

Pepperberg, I. M. (1998). Talking with Alex: Logic and speech in parrots. *Scientific American Presents* 60–65.

Perelman, C. & Olbrechts-Tyteca, L. (1969). *The new rhetoric: A treatise on argumentation.* Notre Dame, IN: University of Notre Dame Press.

Radway, J. A. (1984). *Reading the romance: Women, patriarchy, and popular literature.* Chapel Hill: University of North Carolina Press.

Sivell, S., Elwyn, G., Gaff, C., Clarke, A., Iredale, R., Shaw, C., et al. (2008). How risk is perceived, constructed and interpreted by clients in clinical genetics, and the effects on decision making: Systematic review. *Journal of Genetic Counseling* 17(1): 30–63.

Syed, T., Bölker, M. & Gutmann, M. (2008). Genetic "information" or the indomitability of a persisting scientific metaphor. *Poiesis & Praxis* 5(3/4): 193–209.

Thaler, R. H. & Sunstein, C. R. (2008). *Nudge: Improving decisions about health, wealth, and happiness.* New Haven: Yale University Press.

Venn, C. (2006). The Enlightenment. *Theory, Culture & Society* 23(2): 477–486.

Weaver, R. M. (1965). *The ethics of rhetoric.* Chicago: H. Regnery Co.

Williams, R. (1977/1985). *Marxism and literature.* Oxford: Oxford University Press.

10 An Investigative Bioethics Manifesto

Carl Elliott

Since 1976, the Hastings Center has recognized outstanding work in bioethics with an honor called the Henry Knowles Beecher Award. Beecher was a Harvard anesthesiologist who became famous for his article, "Ethics and Clinical Research," which was published in 1966 in the *New England Journal of Medicine* (Beecher 1966). In that article, Beecher called out his medical colleagues for the mistreatment of human subjects in medical research. The article was not a polemic, or even an editorial opinion. It was a document of wrongdoing. Proceeding in a dispassionate, clinical style, Beecher simply laid out case after case in which medical researchers had violated the rights and trust of their subjects, exposing them to the risk of serious, lasting injury.

"Ethics and Clinical Research" is an unusual article for bioethicists to memorialize. It bears little resemblance to current scholarship in bioethics. Beecher's article was more like a work of investigative journalism, exposing ethical misconduct, or perhaps that of a medical whistleblower, not unlike a tobacco company or pharmaceutical industry insider, calling public attention to wrongdoing in his own field. But the article offered little analysis or interpretation, or even much moral argument. The force of the article came from what it revealed. Beecher was pointing a finger at something that had gone badly wrong. If there was original scholarship in Beecher's article, it came from the legwork of digging up unethical research.

Yet if you look at the scholars after Beecher who have received the Henry Knowles Beecher Award, you will not find anyone who did the kind of work that Beecher himself did. Paul Ramsey, Joseph Fletcher, Hans Jonas, Jay Katz: the list of honored scholars is long and illustrious, but none of its members are known for original investigation. Their work fits much more comfortably within academic departments of philosophy, law, or religious studies. They constructed arguments, developed theories, and wrote policies, but as a rule, they did not dig up dirt and hold it up for public view. Muckraking was not their business. Nor is it the business of most bioethicists today. Investigative reporting is generally considered an activity best left to professional journalists, if it is even considered at all.

In fact, when it comes to investigative journalism, bioethicists appear to have a curious blind spot. The history that bioethicists tell themselves about the field is a roll call of ethical scandals, yet the investigative journalists who brought those scandals to light are rarely mentioned. It is a rare bioethicist who remembers that it was an AP journalist named Jean Heller who first wrote about the Tuskegee syphilis studies for the *Washington Star* (Heller 1972), or that it was Jessica Mitford's article in the *Atlantic Monthly* that set off the debate about drug companies testing drugs on prisoners (Mitford 1973). The relationship of ethicists with journalists is a little more self-serving. The journalists do the grunt work, and then they call bioethicists up for expert commentary. In fact, at the University of Minnesota, where I work, any time you appear in the press this way, the university public relations office will send around a mini news release. The university sees it as great publicity that you were quoted in the *Washington Post* or the *New Yorker*. This is a mark of your importance. But if you actually *write* an article in the *Washington Post* or the *New Yorker*, nobody notices.

Periodically, when the mood of the field turns toward self-examination, bioethicists will debate among themselves the proper relationship between bioethics and the media. Should bioethicists appear on television as experts and pundits? Should they write articles for newspapers and magazines? Occasionally, if ethicists are feeling especially conscience-stricken, they might discuss their roles as whistleblowers, and the circumstances under which they might go to the press if they come across ethical or legal misconduct. What bioethicists rarely consider is the possibility that they might do investigative work themselves.

By "investigative" I do not mean partisan or polemical. Bioethicists have long made it their business to advance substantive moral arguments, often in such exhaustive detail that outsiders are left scratching their heads. Nor am I referring to political activism. There are plenty of bioethicists doing the hard work of raising money for moral and political causes, marching with signs, organizing grassroots communities, and writing opinion pieces. What I mean by investigative is simply a willingness to do the old-fashioned, journalistic task of uncovering facts that have been hidden or obscured. This means talking to sources, reading depositions, examining police records, filing Freedom of Information requests—in general, making it your business to sniff around for wrongdoing. Bioethicists have wide latitude to write about what they choose, and with the tenure code, they have more job protection than journalists do. Why shouldn't there be such a thing as investigative bioethics?

That question has become urgent. In early 2009, the *Atlantic Monthly* published a piece by Michael Hirschorn titled "End Times," which asked: what are we going to do when the *New York Times* is shuttered (Hirschorn 2009)? It was not a frivolous question. The *Seattle Post-Intelligencer* has folded. So has the *Rocky Mountain News*. The *Chicago Tribune* and the *Philadelphia Daily News* are bankrupt, and the *Minneapolis Star-Tribune*, my local paper, has just emerged from bankruptcy. The *Los Angeles Times,* the *San Francisco Chronicle,* the *Boston Globe,* the *Miami Herald,* the *Cleveland Plain Dealer,* the *St Paul Pioneer Press,* the *Detroit News,* the *New York Daily News,* the *Fort Worth Star-Telegram:* all teetering on the brink of collapse. Magazines are not in much better shape. The major American newsweeklies are all struggling; *US News & World Report* has cut back from weekly to monthly publication. Print journalism is collapsing, and many of us are left wondering who, exactly, will report the news.

For at least the past century, and arguably much longer, we have had a media model which involved a devil's handshake: the people who wanted to sell us things footed the bill for our media, and in turn, we tolerated their sales pitch. But that arrangement is disappearing. The digital revolution has decoupled the partnership between mass media and mass marketing. The audience for old media is defecting to the Internet; the content of the Internet is being created by users themselves; and the old distribution channels—like paper, newsrooms, and Hollywood studios—are so expensive that they can't compete. So the old media have tipped over the edge into financial free fall. Newspapers, magazines, television and the music industry are all collapsing into a swirling vortex of devastation, and sucked down with them are the advertisers, who can't persuade anyone to buy ads in media which nobody is using. As Bob Garfield says, "The revolution will not be monetized" (Garfield 2009).

Many people say: good riddance. Cable television has hardly been a healthy influence on American politics, and no one has ever held up the music industry as a model of fair management for struggling musicians. The vast majority of American newspapers do not even come close to the quality of the *New York Times* or the *Washington Post.* For all our mourning over the death of newsprint, it is worth remembering that the pages of most daily newspapers are dominated by advertisements, lifestyle stories, sports reporting, and the comics. Still, if you dig beyond the fluff and filler, you will find a core of important journalistic work that is in danger of disappearing—foreign news coverage, local trench journalism (like city council meetings and school board decisions), and perhaps most alarming of all, investigative reporting, which is expensive to produce and even more expensive to defend against litigation. So the investigative reporters who keep our government honest, who snoop around corporate boardrooms and military installations and the various corridors of political power, looking for muck to rake, are rapidly finding themselves unemployed.

This is all bad news for anyone who cares about medical ethics. The past 20 years have seen some extraordinary changes in medicine: the wholesale movement of clinical research into the private sector, the rise of a multibillion dollar for-profit research ethics review industry, the explosion of direct to consumer advertising of prescription drugs, and a period of unprecedented pharmaceutical industry profits, followed by modest decline. Where there are such extraordinary profits there will also be abuses, many of which have been exposed by investigative journalists, especially at the *New York Times*, the *Los Angeles Times* and the *Wall Street Journal*, which have reported on pharmaceutical ghostwriting, fraud, and off-label marketing. Some impressive investigative reporting on medical issues has also come out of regional newspapers such as the *St. Petersburg Times*, the *St. Paul Pioneer Press*, and the *Milwaukee Journal Sentinel*, among others. But this kind of work is fading by the month.

One of the most remarkable investigative reports to emerge over the past decade came when journalists at *Bloomberg Markets* discovered that a clinical research company called SFBC International had been paying undocumented immigrants to test the safety of new drugs in a dilapidated motel in Miami (Evans et al. 2005). The medical director didn't actually have a medical license; she had a degree from an offshore medical school in the Caribbean. Some of the studies had been given ethics approval by a for-profit ethics committee owned by the wife of a company vice-president. Some subjects told the Bloomberg journalists of alarming effects the studies had on their health. The testing site was not exactly a secret operation. In fact, with 675 beds, the SFBC facility was the largest drug-testing site in North America. SFBC even won awards from major financial magazines for best small business in America. While SFBC was winning awards, however, the motel that served as its primary testing site was so shabby and potentially dangerous that it was eventually demolished by the Miami-Dade Unsafe Structures Board. Almost anybody with a sense of the way clinical trials are supposed to work could have walked through the front door and seen that something was not right. Yet it would never occur to most bioethicists that they should be the ones to do it. Why not?

Of all the ways in which bioethics has changed as a field since the 1970s, the most striking transformation has been its movement into the structures of medical power. Virtually every elite academic health center in America now has a bioethics program, from Harvard and Penn to Johns Hopkins and Stanford. The Institute of Medicine has elected bioethicists to its membership. The National Institutes of Health has established a major bioethics program, as has the Veteran's Administration and the American Medical Association. The Clinton administration was served by the National Bioethics Advisory Commission, the Bush administration by the President's Council on Bioethics, and now the Obama administration by the Presidential Commission for the Study of Bioethical Issues. Even the pharmaceutical and biotechnology industries hire bioethicists now (Elliott 2001; Stolberg 2001).

It was not always this way. Henry Beecher notwithstanding, the scholars that are commonly associated with the early days of bioethics were generally not physicians, nor were they closely affiliated with hospitals, medical schools, or professional medical organizations. These scholars were more likely to be working in divinity schools. Nor were the early institutions of bioethics, such as the Hastings Center and the Kennedy Institute, affiliated with medical institutions. Early bioethics was pragmatic and socially conscious, but it was largely conducted outside academic health centers and professional medical bodies.

In some ways, this transformation of bioethicists from outsiders to medical insiders is unsurprising. Any new field will struggle to achieve social and academic legitimacy, and as bioethics has tried to establish itself as an enterprise worth supporting, it is natural for its practitioners to gravitate towards the people and institutions where power resides. But cozying up to power does not sit easily with the notion of bioethics as a critic of medicine that will keep an eye on its excesses. There are many medical watchdog groups in the U.S., of course—advocacy organizations such as the Public Citizen Health Research Group, Circare, and The Center for Science in the Public Interest, among others—whose purpose is to protect patients or research subjects. Some of these groups, such as Organs Watch, are even based in universities. But these watchdog groups have evolved independently of bioethics, and only rarely have bioethicists been associated with them.

The larger social currents pulling bioethics into medical power structures are complex, but two of them stand out. The first is the emergence of the clinical ethics movement in the late 1980s, establishing ethics consultation as an alternative or supplement to hospital ethics committees. Clinical ethics established an organizational foothold for bioethics in hospitals, and it also provided a way for physicians to defend themselves from the theologians and philosophers who were criticizing medicine from the outside. The second reason behind the transformation was funding from the Human Genome Project, the $3 billion program founded in 1990 which set aside 5% of its budget for ethical and legal issues. The ELSI program of the Human Genome Project funneled an unprecedented amount of research funding into bioethics, making it possible for academic health centers to establish bioethics programs using the soft-money funding models familiar from other kinds of medical research institutes. This ELSI funding has turned out to be a mixed blessing. More bioethics programs have been established, but many more bioethicists now depend on securing research grant money in order to keep their jobs. This is not a recipe for vigorous dissent.

Nor is dissent what it has produced. In the early days of bioethics, the mood of the field was generally skeptical of medical authority, concerned over the research imperative, and suspicious of any reflexive trust in medical progress. This skeptically-minded scholarship has not gone away, but during the late 1990s and 2000s it was joined by work with a more celebratory tone. This was a period in which a number of bioethicists argued vigorously for the use of medical technology not merely to cure illness, but to enhance the capacities of healthy people. Some ethicists argued that every citizen had an obligation to participate in medical research, or even to genetically enhance their children. Others began to reconceptualize medical research, transforming it from a source of potential harm or exploitation to a social benefit which could be unjustly denied to women and minorities. Many ethicists began to press for more embryonic stem cell research. Others made wildly optimistic predictions about the future of medicine. In 2003, *American Journal of Bioethics* editor Glenn McGee told the *Philadelphia Inquirer* that within five years, American would be genetically testing themselves at home with iPod-sized devices which would allow drugs to be tailor-made for their personal genomes (Lotozo 2003).

Bioethicists also began to establish a presence in the private sector. Some bioethicists joined advisory boards set up by biotechnology companies such as Geron, Advanced Cell Technologies, and DNA Sciences. Others went to work for commercial Institutional Review Boards, which review research studies in exchange for a fee (Elliott & Lemmens 2005). A handful of ethicists set up their own businesses, such as Glenn McGee's Bioethics Educational Network or Bruce Weinstein's Ethics at Work. As the pharmaceutical industry boomed, building itself into the world's most profitable industry, a number of prominent ethicists began consulting for it as well: Tom Beauchamp and Robert Levine for Eli Lilly, Jonathan Moreno for GlaxoSmithKline, James Childress for Johnson & Johnson, and Arthur Caplan for Pfizer, Celera, Aventis, AstraZeneca, Wyeth, Dupont, Monsanto, and Merck.[1] In 2003, Eli Lilly began marketing its enormously expensive sepsis drug, Xigris, with a public relations campaign called "The Ethics, the Urgency and the Potential," which was capped off with $1.8 million dollar bioethics task force (Regalado 2003).

Perhaps the most striking example of the way bioethics was transforming itself came when an 18-year-old Jesse Gelsinger died in a gene therapy study at the University of Pennsylvania in 1999 (Stolberg 1999). Gelsinger had been born with ornithine transcarbamylase (OTC) deficiency disorder, a rare metabolic illness which kills most children shortly after birth. Gelsinger, however, had a mild form of OTC disorder which he had been able to control with drugs and diet. He was a relatively healthy young man. But when the Pennsylvania researchers injected Gelsinger with a modified cold virus, intended to carry corrective genes to the liver, he rapidly went into multisystem organ failure. He died within days.

Gelsinger's death stunned many scientists, who had seen gene therapy as an extraordinarily promising area of research. It also shocked bioethicists. Paul Gelsinger, Jesse's father, sued the University of Pennsylvania, charging that the Penn researchers had not been completely honest about the potential risks of the study. An FDA investigation found that Gelsinger should never have been considered for the trial because of the condition of his liver (Stolberg 1999a & b). The study was also compromised by serious conflicts of interest. James Wilson, the researcher in charge of the study, had held a 30% controlling interest in Genovo, the biotechnology company which stood to profit from the study. Genovo also provided $4 million a year to the research program Wilson directed, the Institute for Human

Gene Therapy. Still another ethical controversy concerned the design of the trial. Most risky trials of new drugs for serious illnesses are done on the patients most severely affected with the disease, and whose prognosis is very poor without any treatment. But the Penn researchers had decided to test the gene therapy on adult patients, which meant that relatively healthy patients such as Jesse Gelsinger were exposed to the risk of serious injury and death.

Shortly after Gelsinger died, two bioethicists at the University of Pennsylvania, Arthur Caplan and David Magnus, published a newspaper op-ed about the study. They did not criticize the study, however. They did not even mention its ethical problems. Rather, they worried that Gelsinger's death might lead to more regulation and slow the pace of medical research. "We do a disservice to Jesse Gelsinger and others who have been hurt or killed in medical research by simply adding layers of bureaucracy in the path of clinical research," Caplan and Magnus wrote. "If we are not careful we may wind up allowing our collective grief over the death of a young research subject to justify the imposition of bad public policy governing the future of gene therapy" (Caplan & Magnus 1999).

The article by Caplan and Magnus was not widely noticed, but it suggested a new role for bioethicists: defenders of their employers in the face of scandal. Other ethicists have followed suit. When a conflict of interest controversy erupted at the NIH, revealing that many government-employed scientists were being paid by the pharmaceutical industry and failing to report the income, it was the director of the bioethics program who led efforts to fight reform (Wadman 2005; Weiss 2005). When the Columbia space shuttle exploded in 2003, raising questions about the priorities of NASA and the future of the shuttle program, an editorial appeared by NASA's chief bioethicist, praising the spirit of space discovery (Wolpe 2003). When a ghostwriting scandal erupted at McGill University in 2009, revealing that a McGill researcher had signed onto an article about hormone replacement therapy produced by Wyeth medical writers, it was a prominent McGill ethicist who came to the rescue in the *Montreal Gazette* (Somerville 2009).

Of course, this kind of public defense does not necessarily mean that ethicists are compromising their principles, or bowing to internal pressure, or even consciously taking up arms on the side of their employers. For some, speaking out publicly may even be a matter of conscience. But the fact that the moral convictions of bioethicists line up perfectly with the mission of the medical-industrial complex does not make the problem any less troubling. In fact, it may even be more troubling than if bioethicists were being pressured. A true believer can be more dangerous than a captive.

About 10 years ago, not too long after I started teaching at the University of Minnesota, I got an unusual phone call from a psychiatrist named Faruk Abuzzahab. Abuzzahab once been a member of our Department of Psychiatry, and he wondered if he could sit in on a medical ethics class I was teaching. Apparently there had been some trouble in a research study Abuzzahab had done, and he was being disciplined by the state licensing board. A course in medical ethics was part of his punishment.

Abuzzahab seemed pleasant enough on the phone. He even told me that he had been chair of the ethics committee for the Minnesota Psychiatric Society. Against my better judgment, I agreed to let him audit the class. Abuzzahab was a quiet but cooperative student, and I had nearly forgotten about him when a for-profit clinical trials site called Prism Research opened up in St. Paul. When I looked at the Prism website, I saw that Abuzzahab was listed as one of their clinical investigators. I began to wonder: what exactly was the incident that had brought him to my class?

As it turned out, Abuzzahab's medical license had been suspended for two years. The state licensing board had judged him responsible for the deaths and injuries of 46 people under his care, several of whom had committed suicide. Seventeen of these people had been research subjects in studies Abuzzahab was doing for the drug industry. Abuzzahab had been taking patients with severe mental illnesses and cycling them in and out of studies of new unapproved drugs. Many subjects did not meet eligibility criteria, and often Abuzzahab kept subjects in the studies even after they got even worse. One of the subjects, Susan Endersbe, had been a patient at Fairview Hospital, a teaching hospital for the University of Minnesota. She had a diagnosis of schizophrenia and had attempted suicide. When Abuzzahab gave her a day pass to leave the hospital unaccompanied, she threw herself into the Mississippi River and drowned.

The backdrop to Abuzzahab's story is the transformation of the clinical trials industry in the 1990s. Beginning in the early 1990s, the pharmaceutical industry began to move clinical trials out of universities and into the private sector. By the mid-2000s, clinical trials were largely managed by Contract Research Organizations, conducted by private physicians, and overseen by commercial Institutional Review Boards, each of which operates as a for-profit business. Many ethicists simply assume that the regulatory apparatus set into motion during the 1970s to oversee trials has followed those trials as they have moved out of universities and into the private sector. Today, however, clinical research is such a vast multinational industry that it far exceeds the capacities of those oversight bodies. In the years after Abuzzahab's suspension, for example—and subsequent rehabilitation by virtue of my class—Abuzzahab suffered very little professional setback. In 2003 he was even awarded a Distinguished Life Fellowship by the American Psychiatric Association.

I eventually wrote about Abuzzahab in an article on the clinical trials industry, which was published in the *New Yorker* in 2008 (Elliott 2008). Looking back at the episode, however, what strikes me is how little responsibility I originally felt to check into Abuzzahab's past. Almost 10 years passed before I published the article. During that time, Abuzzahab worked for at least a dozen different pharmaceutical companies, either giving talks or conducting clinical trials. At the time of that first phone call, however, it simply did not occur to me that I might be equipped to write an investigative report.

Investigative reporting is not for everyone, of course. Some people are not suited for it; others are just not interested. And it does require a different kind of work than most bioethicists are accustomed to. Even the relatively modest reporting I have attempted has required a lot of help from seasoned experts willing to help me navigate FDA websites, search for financial filings, examine court depositions, and find out which attorneys are willing to share unsealed legal documents. I would not want to suggest that investigation should replace argument or analysis. I merely want to suggest that it should stand alongside the other things that bioethicists do.

Investigative bioethics would face some significant institutional hurdles, of course. For example, how would promotion and tenure committees look at this kind of work? While academics generally do not expect to be paid for their writing, they are unlikely to invest much time and effort into work that will not translate into academic advancement. Nor is it clear exactly where investigative bioethics could be published. Most academic journals are not equipped to evaluate works of journalism, much less to employ the rigorous fact-checking procedures of magazines such as the *New Yorker* and the *Atlantic Monthly*. And while a handful of academic journals (such as the *BMJ*) occasionally publish investigative journalism, most investigative work has traditionally been published in the very magazines and newspapers that are now facing such financial trouble, courtesy of the digital revolution.

In other ways, however, the time for investigative bioethics could not be better. While it may be true that the digital revolution is killing investigative journalism *jobs,* it is creating investigative journalism *resources.* The web has made it possible for ordinary people to investigate wrongdoing in a way that would have been unthinkable even five or six years ago. Court documents, disciplinary files, corporate memos, financial reports: these things are available to virtually anyone now, not just trained reporters with institutional backing. Take, for instance, the pharmaceutical industry, a notoriously secretive business which has suddenly become much easier to investigate. At the University of California San Francisco, a group of former expert witnesses have created the Drug Industry Document Archive: a database of documents from litigation against the drug industry. It has confidential memos, ghostwritten journal articles, marketing plans, contracts, and billing records, all of it searchable by keyword online (Drug Industry Document Archive 2005). In Minnesota, legislation has mandated a pharmaceutical industry sunshine law, which requires the drug industry to report any gift or payment to a Minnesota physician over $100. A searchable database allows anyone with a computer to find out how much money a given physician has received (Twin Cities.com Pioneer Press Archive 2008). Several pharmaceutical companies have followed suit, posting their payments to physicians and organizations online. And those are only a few examples from a rapidly expanding pool of information. A cadre of bloggers, reporters, activists, and attorneys now post leaked or unsealed court documents on their websites, often with commentary and analysis.

New avenues for investigative reporting are emerging as well. Some of the most promising avenues are the non-profit news organizations dedicated to reporting in the public interest. Some of these groups—such as the Center for Investigative Reporting in San Francisco and the Center for Public Integrity in Washington, D.C.—have been around for many years, but are now attracting renewed attention as a result of the journalism crisis. Other groups are relatively new. Pro Publica, for example, was established in 2007 and currently employs 32 journalists who give their stories away for publication in news outlets such as *60 Minutes* and the *New York Times*. Some investigative work even is being done in universities. At Northwestern University, a partnership has been established between the Medill School of Journalism and the Innocence Project, a non-profit organization which uses DNA evidence to exonerate prisoners who have been wrongfully convicted. In 2008, American University established the Investigative Reporting Workshop in the School of Communication, led by Charles Lewis, the founder of the Center for Public Integrity.

Investigative bioethics also has some more traditional academic precedents. Medical anthropologists and sociologists have been doing ethnography for years, much of it with a sharply critical edge. For example, books such as Roberto Abadie's *The Professional Guinea Pig* (2010), Adriana Petryna's *When Experiments Travel* (2009), and Jill Fisher's *Medical Research for Hire* (2008) have exposed a side of the clinical trials industry that outsiders rarely see. Of course, ethnography differs from investigative reporting in that its purpose is not to expose wrongdoing. But some recent ethnography has come very close to investigative reporting—most notably, the work of Nancy Scheper-Hughes on the global traffic in human organs. But perhaps the clearest example of investigative bioethics is the work of Steve Miles on torture, especially his book *Oath Betrayed* (2009). Drawing on autopsy reports, prison medical records, FBI notes, and other documents obtained through the Freedom of Information Act, Miles documents the complicity of military medical personnel in the abuse of detainees during the Bush administration's war on terror. His work is as close to the spirit of Henry Beecher's article as anything in bioethics today.

A few years ago, at a national bioethics conference I attended, the director of a bioethics center stood up and repeated what a donor had said to her: "Your field is full of charlatans and scam artists, and I want to know what you plan to do about it." The donor may have been overstating the case, but there are many people who would agree with him, even within bioethics itself. It is not easy to think of many other academic fields with this sort of reputation. The usual rap on academics is that they are disengaged with life outside the university, hopelessly lost in a world of books and lectures. The rap on bioethicists is just the opposite: engaged with the broader culture, but engaged in a way that is self-serving and even harmful. "Fee-for-service philosophers" is the term used by Ruth Shalit in the *New Republic* (Shalit 1997). The sociologist Jonathan Imber summed up the prevailing view 10 years ago when he called bioethics "the public relations division of modern medicine" (Imber 1998).

It is striking just how many notable bioethical scandals of the past 20 years or so have occurred at universities with well-established bioethics programs: Penn, Duke, Johns Hopkins, and Toronto, among others. The University of Minnesota, where I work, and whose Center for Bioethics will soon celebrate its 25th anniversary, has been host to a string of ethical embarrassments over the past 15 years, beginning with the anti-lymphocyte globulin (ALG) scandal uncovered in the early 1990s, continuing on through a series of conflict of interest scandals involving the Departments of Orthopedic Surgery and Psychiatry, a case of scientific fraud in the Stem Cell Institute, a double-dipping investigation in the School of Public Health, and most disturbing of all, a suicide in an industry-funded drug trial in the Department of Psychiatry.[2]

It would be unfair to blame bioethicists for scandals in which they had no involvement, of course. Simply working at a scandal-ridden institution does not mean culpability. But it would also be unfair to overlook the fact that a large part of the reason why bioethics has risen to prominence as a field is the notion that it can help remedy or prevent ethical wrongdoing. Many people believe that the establishment of more bioethics centers, more bioethics courses, more bioethical guidelines, and policies and taskforces will actually make the world of medicine a better place. It would be disturbing to think that bioethics might have the opposite effect.

NOTES

1. For financial disclosures by Beauchamp and Levine, see Beauchamp et al. (2002) and Eli Lilly and Company Corporate Citizenship Report 2005–2006, p. 16. For Moreno, see the home page at the University of Pennsylvania (*http://hss.sas.upenn.edu/mt-static/faculty/department_faculty/jonathan_moreno_phd_professor.php*). For Childress, see Annual Report, Center for Biomedical Ethics, University of Virginia July 2002–June 2003, p. 13. For Caplan, see the expert testimony submitted in In Re Diet Drugs (Phentermine/Fenfluramine/Dexfenfluramine) Products Liability Litigation; Sheila Brown, Sharon Gaddie, Vivian Naugle, Quintin Layer, and Joby Jackson Reid, individually and all others similarly situated. Plaintiffs v American Home Products. Civil Action No. 99–20593. In the United States District Court for the Eastern District of Pennsylvania. See page 100 of Caplan's CV.
2. For a sampling of these scandals at the University of Minnesota, see: Art Hughes, "U of M Medical School Still Recovering 10 Years after ALG scandal," Minnesota Public Radio, February 28, 2006; available from <http://minnesota.publicradio.org/display/web/2006/02/28/uofmhospital/>; Janet Moore, "U Surgeon's Deals with Medtronic Draw Fire," *Minneapolis Star Tribune*, July 29, 2009; Tony Kennedy, "Two U profs suspected of double-dipping,"

Minneapolis Star Tribune, April 20, 2008; Nicholas Wade, "Panel Finds Flawed Data in a Major Stem Cell Report," *New York Times,* February 28, 2007; Maura Lerner, Josephine Marcotty and Janet Moore, "U doctor on ethics panel was disciplined," *Minneapolis Star Tribune,* December 21, 2008; Jeremy Olson and Paul Tosto, "Dan Markingson had delusions. His mother feared that the worst would happen. Then it did," *St. Paul Pioneer Press,* May 23, 2008.

REFERENCES

Abadie, R. (2010) *The professional guinea pig: Big pharma and the risky world of human subjects.* Durham, NC: Duke University Press.

Beauchamp, T., Jennings, B., Kinney, E., and Levine B. (2002). Pharmaceutical research involving the homeless. *Journal of Medicine and Philosophy* 27: 547–564.

Beecher, H. (1966). Ethics and clinical research. *The New England Journal of Medicine* 274: 1354–1360.

Caplan, A. & Magnus, D. (1999). Overregulating research. *Chicago Tribune,* December 21, 1999. 31

Drug Industry Document Archive (2005). Accessed online at http://dida.library. ucsf.edu/. January 2011.

Elliott, C. (2001). Pharma buys a conscience. *The American Prospect* 12: 16–20.

Elliott, C. (2003). Not-so-public relations: How the drug industry is branding bioethics. *Slate,* December 15, 2003.

Elliott, C. (2008). Guinea-pigging. *New Yorker,* January 7, 2008. pp. 36–41.

Elliott, C. & Lemmens, T (2005). Ethics for sale: For-profit ethical review, coming to a clinical trial near you. *Slate,* December 13, 2005.

Evans, D., Smith, M. & Willen, L. (2005). Big pharma's shameful secret. *Bloomberg Markets* 14: 36–62.

Fisher, J. (2008). *Medical research for hire: The political economy of pharmaceutical clinical trials.* Piscataway, NJ: Rutgers University Press.

Garfield, B. (2009). *The chaos scenario: Amid the ruins of mass media.* Nashville, TN: Stielstra Publishing.

Heller, J. (1972). Syphilis patients died untreated. *The Washington Evening Star,* July 25, p. 1.

Hirschorn, M. (2009). End times. *Atlantic Monthly,* January/February, 2009. pp. 41–45.

Imber, J. (1998). Medical publicity before bioethics: Nineteenth-century illustrations of twentieth-century dilemmas. In R. DeVries and J. Subedi (Eds.), *Bioethics and society: Constructing the ethical enterprise.* Upper Saddle River, NJ: Prentice-Hall. 16–37.

Lotozo, E, (2003). Bioethicist foresees a wild frontier in genetics field. *Philadelphia Inquirer,* November 2, 2003, 6C.

Miles, S. (2009). *Oath betrayed: America's torture doctors.* Berkley, CA: University of California Press.

Mitford, J. (1973). Experiments behind bars: Doctors, drug companies, and prisoners. *Atlantic Monthly* 23: 64–73.

Petryna, A. (2009). *When experiments travel: Clinical trials and the global search for subjects.* Princeton: Princeton University Press.

Regalado, A. (2003). To sell pricey drug, Eli Lilly fuels debate over rationing. *Wall Street Journal,* September 18, 2003, p. A1.

Shalit, R. (1997). When we were philosopher kings. *New Republic,* April 28, p. 24.

Somerville, M. (2009). Ethics, like law and science, evolves over time. *Montreal Gazette*, August 28, 2009. p. A15.

Stolberg, S. G. (1999a). F.D.A. officials fault Penn team in gene therapy death. *New York Times*, December 9, 1999. Available from <http://www.nytimes.com/1999/12/09/us/fda-officials-fault-penn-team-in-gene-therapy-death.html?ref=sherylgaystolberg> Accessed 13 June 2011.

Stolberg, S. G. (1999b). The biotech death of Jesse Gelsinger. *New York Times Magazine*, November 28, 1999. Available from <<http://www.nytimes.com/1999/11/28/magazine/the-biotech-death-of-jesse-gelsinger.html?ref=sherylgaystolsberg&pagewanted=print> Accessed 13 June 2011.

Stolberg, S. G. (2001). Bioethicists fall under familiar scrutiny. *New York Times*, August 2, 2001, p. 1.

TwinCities.comPioneerPressArchive(2008).Availablefrom<http://extra.twincities.com/CAR/doctors/> Accessed 10 January 2011.

Wadman, M. (2005). NIH workers see red over revised rules for conflict of interest. *Nature* 434: 3–4.

Weiss, R. (2005). NIH workers angered by new ethics rules. *Washington Post*, Thursday, February 3, 2005: A25.

Wolpe, P. R. (2003). Is space program ethical? NASA's first bioethicist tells how he decided. *Philadelphia Inquirer*, February 5, 2003. Available from: <http://articles.philly.com/2003-02-05/news/25451412_1_space-exploration-space-program-manned-space> Accessed 13 June 2011.

11 The Question of "the Public"

Christian O. Lundberg and Ross Smith[1]

In this essay we reconstruct what we take to be the primary stasis points that unite this volume's diverse set of essays in order to come to judgments, provisional conclusions, and potential directions for the intersection of scholarship in bioethics and communication studies. The term "stasis" originally appeared in Greek antiquity with the negative connotations of "faction" and "civic discord." In Aristotle's hands, however, stasis becomes a productive site for thinking through exactly what is at stake in a given argument (Kalimtzis 2000, p. xii). The stasis points that we explore revolve around the themes suggested by the volume's title: "Bioethics, Public Moral Argument and Social Responsibility." We do not take up the contested term "bioethics" here in any detail. Instead we focus on the points of stasis present in the idea of the "the public," the articulation of publics and "moral argument," and, extending from the preceding two stasis points, the concept of "social responsibility." In defining and engaging these major stasis points we intend to point toward three basic conclusions, demonstrating: 1) the need to understand a public as something that is not made in advance, but rather is constituted in the process of argument, coupled with a moral claim for building public capacity for argument around bioethics; 2) the necessity of thinking about a public in relation to the unique capacities for attention and modes of rhetorical mediation that constitute it as a democratic form; and 3) the imperative to understand the interdependence of claims surrounding bioethics controversies and the processes of moral argument that mediate them as a mode of instantiating social responsibility.

STASIS POINTS

The Public

Arthur Strum observed that the concept of "the public" has lost much of the conceptual force and theoretical specificity that once made it useful for analytical work. Conceptions of the public have suffered this fate at the hands of "loudly competing but mutually incomprehensible disciplinary . . . spheres, each with its own practical presuppositions and theoretical-methodological languages" (Strum 1994, p. 163). Though perhaps overly pessimistic, Strum's diagnosis points to a central difficulty in thinking through the problem of the public. When invoked without any explicit definition, the idea of the public often contains a number of potential meanings, including but not limited to: a space for actions that are seen by and effect others (the public sphere); a modifier that names a specific quality of an issue at hand (the public interest); a specific audience that is addressed (a public); or the mass of all possible readers or consumers of a text (the general public). Sorting out which understanding is operative in any given iteration of the term "public" has significant implications, particularly in relation to understanding a moral argument as "public."

No essay in this volume explicitly defines the imagined character of the public, but there are some suggestive hints. Carl Elliott, for example, argues that academic bioethics has basically become parasitic on the pharmaceutical industry, too often relying on industry funds, so that there are probably more ethicists consulting for the industry than criticizing it (Elliott, this volume). He thus claims that bioethics scholars fail to protect *the public interest* by serving as critics of the pharmaceutical establishment because of a structural conflict of interest between the funding that supports their work (both directly and by providing funding to the academic medical centers in which they work) and the modes of criticism that their scholarship should ideally produce. This difficult institutional bind for the critical impulses of academic bioethics is compounded, according to Elliott, by journalistic failures in reporting on bioethics. This failure is not only specific to reporting on bioethics, which involves difficult issues that often hinder public engagement; it is abetted by a more general decline in the fortunes of print journalism. In this context, one of the most critical traditional mediums for holding the medical-industrial complex accountable for its actions will probably be increasingly impotent in bringing its ethical lapses to light.

Elliott urges that the solution to this dual problem is for bioethics scholars to take up the practice of investigative bioethics. He claims that investigative bioethics might be most effective in doing the "old-fashioned leg-work" of going out and "sniffing around." The primary goal of this task is not to make the kind of arguments usually directed at fellow academics about the ethics of a specific practice, but rather to write compelling narratives about abuses of human subjects and the conflicts of interest that help to produce them.

What is the conception of the public latent in this call for investigative bioethics? It is difficult to disagree with Elliott's arguments that bioethics scholars ought not to be beholden to the medical-industrial complex and that someone needs to take up the standard for informing the public about the practices of the biotechnology and pharmaceutical industries. But to what extent can we expect this practice to be effective? There are two versions of the public present here that are often understood to be virtually interchangeable. One understanding of the public relates to the question of public interest, and one relates to the public as a mass audience.

The first account of the public holds that good journalistic practice necessarily serves the public interest because it makes the public aware of issues and adds to the marketplace of ideas. To put something in print by its very nature adds to the public's capacity to understand, make informed judgments, and ultimately act on the information that it is provided. The second understanding of the public operative here takes the article "the" in the phrase "the public" quite seriously. Here the public is understood as an undifferentiated mass of media consumers, composed of individuals, not specifically marked by either social or ideological positions and equipped with the means, attention span, and capacity to consume information in a way that advances a social cause. In order for both journalism and investigative bioethics to have any political effect, one must presume that one or both of these framings of the public is operative.

How can the public effectively be engaged in discourse about moral and policy controversies in biotechnology—that is, about issues that involve an intersection between science and ethics? If the most important role of the media in the bioethics debate is to accurately and fairly report and interpret the facts, then reporters need basic competency in biological sciences to ask the right questions, and researchers need the translational tools to make themselves clear to reporters.

Many have argued recently, however, that the norms for reporting have changed. Instead of emphasizing factual clarity, journalists have increasingly interjected their own voices into reporting. This change in reporting could be described as failing the public, by requiring them to determine which claims are accurate and which claims reflect an agenda—precisely what good reporting is supposed to protect against. The deliberate or unreflective insertion of journalistic voice into the public dialogue could be viewed as distorting the public debate and thus in turn demanding an ethic of reporting that sets aside personal bias in the realm of public discourse, so that reporting can be transparently truthful.

It is difficult to take issue with the argument that reporters ought not misrepresent the facts in the name of pursuing an agenda, or that they should not present their own dispositions on an issue as facts. At the same time, it is also possible to ask whether anything in public discourse is ever transparently truthful or if facts can ever be neutral. Every fact is presented to an audience for a specific purpose, is tied to the narrative context within which it is presented, and is always value-laden. This "fact" regarding "the facts" provides an important caveat to concern about requiring the public to evaluate claims in areas in which they have no expertise; the public is inevitably in this position. Thus, we ought not give up on an ethic of objectivity in journalism, but in addition must recognize that it is essential to help develop in the public the capacity not only for sorting facts from facts, but also for sorting bias from objectivity and ultimately for evaluating the strength of arguments as arguments.

This friendly supplement to efforts to change either reporting or who reports argues that the most important thing is not the production of information, but the character of the public who consume it. It is critical to recognize that "the public" is not a pre-constituted and uniform entity waiting to be informed. The idea of the public as a pre-existing mass composed of all possible consumers is untenable in the light of the economies of attention and circulation that produces publics. Michael Warner has argued that much of our understanding of the public, especially as conceived in theories of mass media, is informed by an ideology of universal publicness that relies on the idea that the public is composed of an audience of all potential listeners (Warner 2002). Warner identifies this notion of "the public" as an ideological fiction. Working from the presupposition that "the" public is a pre-existent social form elides the substantial diversities that make up actually existing publics, including forms of diversity present in differing identities and subject positions, capacities for public deliberation and involvement, and disparate economies of attention. In place of what he identifies as a "hegemonic" conception of the public, Warner follows critics such as Nancy Fraser in arguing for a conception of individual "publics" as an alternative to a vision of the mass public (Fraser 1990).

Publics are not given in advance; rather, they are made by concrete moments of attention to the circulation of texts. Here a public emerges when subjects who are otherwise strangers to each other pay attention to a shared text. We should not think of the public as a mass eagerly awaiting each successive news story or critical piece on practices in bioethics: instead, text-based arguments about bioethics issues only succeed insofar as they cobble a public of attention together around an issue or a story. The task for reporting and critical commentary on bioethics issues is thus to constitute a public around an issue, not to inform "the public" of an issue that is presumed to be in its interest in advance.

As an extension of the thesis that publics are products of attention, we ought to pay heed to the changing modes of mediation that constitute public experience (Warner 2002). The decline in print journalism may be seamlessly supplemented with blogging and other forms of electronic journalism; even so, we ought to keep in mind Marshall McLuhan's argument that the "medium is the message" (McLuhan 1994). This means that thinking about the public must take into account the sea-change in practices of public attention and action and modes of mediated consumption that the new media environment is bringing about. The everyday life of publics is informed by a much wider and arguably much more intense range of meaning-making media than print and electronic journalism. Entertainment media, new forms of online interaction, and other emergent modes of media consumption all play a role in how we position ourselves relative to others, and how we understand our relationship to the significant questions that inform bioethical inquiry, such as the meaning of the person, the possibilities for deliberative decision-making, and the character of human agency. It is essential to consider the role that new media themselves play in constituting humans as ethical actors.[2]

The problem of public media practices is the point where Roxanne Parrot's reflections on the difficulties of health communication (Parrott, this volume) become most salient, particularly as they relate to the dilemmas of translational research—a task which by its very nature implies a whole set of communicative difficulties in engaging publics around the question of bioethics. Parrot identifies a concern that the public lacks capacities for dealing with the technological complexities of bioethics; as she frames it, "scientific illiteracy" is a barrier to "translational research." Parrot also argues that the general capacities for public literacy that might make public argument over bioethics efficacious are also on the decline. She identifies two forms of "media illiteracy": "content inaccuracies" and a "lack of detail" in media presentations of issues relating to bioethics. What is perhaps most important about Parrot's argument is the "lack of detail" hypothesis. Parrot's treatment of this problem involves a number of issues that distort media treatments of bioethics issues or aim them at a lowest common denominator in public discourse, including a bias for recency, profit motives for "negative reporting," and the significant effects of often inaccurate entertainment media on public understandings of issues in bioethics. Parrot pays close attention to the politics of media technologies and the institutions that employ them in driving public knowledge about issues of bioethical concern—a useful supplement for gleaning a fuller picture.

Further, presuming that the mere presentation of good information or even the uncovering of some egregious moral wrong will result in public action ignores the reality that public capacities for reading, deliberation, and response are uneven. Acknowledging this unevenness is a prerequisite to effective public interventions in bioethics, whether by scholarship or by journalism, because this realization drives us toward the necessity of cultivating public capacities for deliberation and response. John Dewey, in *The Public and its Problems* (Dewey 1927) makes exactly this case in response to claims like Walter Lippmann's in *The Phantom Public* that the idea of reasoned public deliberation was an "unattainable ideal, bad only in the sense that it is bad for a fat man to try to be a ballet dancer" (Lippmann 2005, p. 29). Dewey argued that we ought to redouble our efforts in education, since the alternative to an engaged public was the free reign of technological production without any regulating principle. The debate in the 1920s parallels almost exactly the claims made in these essays that contemporary innovations in biosciences are rapidly outpacing the ability of the general public to engage them. For us, as for Dewey, the only alternative is to cultivate in publics a capacity for engaging the controversies surrounding the biosciences. Without a strengthened capacity for deliberation in general, and without the basic knowledge of the context of biotechno-logical controversies, we are at risk of losing any possibility for democratic input into the conduct of the biomedical sciences. Thus, there is a strong complementarity between the specific argumentative claims embodied in public interventions in bioethics and the everyday work of communica-tion studies, which takes as its pedagogical object the cultivation of gen-eral capacities for public discourse. It almost goes without saying here that strengthening a commitment to the skills of debate and deliberation is an important precursor to engaging the public in controversies in bioethics.

The question is whether or not bioethics and communication scholars can work effectively to build the capacities for deliberation and advocacy that serve as preconditions for the success of any deliberative process. In the absence of such capacity, the illusion of "factual" reporting, or of an insti-tutional commitment by bioethics scholars to protecting the public interest, may lend unwarranted democratic legitimacy to bioscientific practices by inducing public complacency. To repeat: capacity-building is practiced not by informing a public that exists in advance, but rather by building a spe-cific public, endowed with specifically public capacities, around the issues of bioethics.

PUBLIC MORAL ARGUMENT AND SOCIAL RESPONSIBILITY

What kind of capacities? Certainly it is important to cultivate informa-tional and deliberative capacities, but since no communication occurs without some value orientation, we also need to cultivate capacities for moral argument. Capacity-building requires attention to the communica-tive framework for bioethical argument, and to the capacities necessary for advocating moral arguments and making judgments between competing moral claims. David Zarefsky's and Jonathan Moreno's contributions to this volume move us a long way toward establishing such a framework. Both Zarersky and Moreno argue for a conception of moral argument that is pragmatic, pluralist, and focused on the legitimating function of moral deliberation in a democracy. Zarefsky argues that democracy implies the simultaneous necessity of majority rule and of heeding minority opinions. The result is a commitment to argumentative deliberation between the majority and minorities in a fully functional democratic culture. Citing Lincoln, Zarefsky claims that the essence of moral argument in a demo-cratic context is embodied in a moral commitment to "seek out the assent of our fellow human beings" through deliberative exchange (Zarefsky, this volume, pp. 4–5).

This commitment positions argument as the *sine qua non* of democ-racy. Democratic argument is thus always rooted in an appeal to *phrone-sis* or prudential judgment, bounded by the strength of the evidence rather than the degree of intensity or moral certainty that underlies it. Zarefsky emphasizes a constant and productive tension between the majoritarian and prophetic voices in a democratic polity. The prophetic voice, which stands in a position of relative marginality to majority rule, both calls for a different ordering of things and holds the majority accountable for living out its ostensible moral commitments. This is not a contest over abstract moral goods. Instead, moral decision-making is always bound not only by the context within which moral disputes arise, but also by the particular case that calls for a moral judgment. Framing democratic moral argument as always bound to the case acknowledges moral calculations as local and provisional to a given dispute, and as governed by a process of policymaking, as opposed to the application of absolute *a priori* rules. Though public or democratic moral argument cannot be decided in advance by a set of context-independent moral rules, Zarefsky indentifies six common strategies of moral argument. Though advocates may have differing conceptions of what action ought to be taken in a given case, the presence of such common strategies reveals substantial ground for shared and potentially consensual moral argumentation, even in the context of difficult antagonisms.

Arguing in a similar thread, Moreno (this volume, p. 23) takes up the question whether bioethics is "an ideology masquerading as moral philosophy." Moreno begins by noting that there is a paradox in consensus. The democratic ethos upholds the value of consensually-based decision-making, but the judgments of the demos are notoriously fickle. While noting that a degree of moral heroism is latent in the history of bioethics, Moreno warns against the all too easy valorization of individual judgment at the expense of "deep consensus." Deep consensus is distinct from both superficial consensus (which only seeks after either coerced or trivial points of agreement) and compromise (which necessarily entails the dilution of ethical commitment). Deep consensus must respect both the outcomes that result from it (product) and the commonality of means by which it is achieved (process).

Deep consensus must paradoxically be both communally produced and constantly revisable in the face of argument. Both in Moreno's argument for pragmatic naturalism and Zarefsky's argument for a necessary and productive tension between the prophetic and majoritarian voices in democracy, we have returned implicitly to the question of publics in public moral argument. To achieve deep consensus, interlocutors must attend to the particular context within which they argue, and they must be endowed with the requisite capacities for sensitivity toward local argument practices that are the condition of possibility of deep consensus or productive tension between the prophetic and majority voice. Moral judgments ought be both revisable and argumentatively responsive to the context in which they are put forward.

So far, we have mostly addressed the issue of capacities for public deliberation from the perspective of argumentative invention and the modes of relation to audiences that moral argument demands if it is in fact to be truly public. But what we are ultimately after is communicative ethics. Communication is a multifaceted phenomenon that must be responsive to the context that produces it, which is a form of relation to the other that implies a notion of social responsibility and entails commitments to provisional and revisable conclusions. As soon as we identify such a radically contextual and communal conception of moral meaning-making, then it will not do to simply say that interlocutors ought to listen to and be affected by the arguments that others make in public. What is needed is also a way of addressing the capacities for listening to and interpreting the arguments that other advocates make in a democratic context.

Three essays help provide a richer conception of this listening process: Larry Churchill's reflections on the relationship between determinism and agency in religious discourses; Rebecca Dresser's defense of dignity as a contextually embodied practice that is not reducible to autonomy; and Celeste Condit's incisive argument for considering affect in thinking through biomedical discourses.

One might take Churchill's essay as an exemplar of a mode of rhetorical interpretation that would aid in building capacities for public moral argument. Churchill notes that certain discourses of moral determinism seem to deny moral agency. For example, claims by religiously devout persons that a specific medical outcome is determined by fate appear to minimize an actor's own moral agency in deciding a course of action—such as in a decision for or against medical treatment near the end of life. Yet Churchill argues that claims about "fate" or destiny must be understood through their rhetorical function in a given community, which might produce an understanding quite different from how one they might be understood by those outside that public. Thus he posits (this volume, p. 84):

> that the prevalent popular, historic, and clinical assumptions about genetics often verge on determinism, [and] that religious/spiritual interpretations of determinism, expressed as "fate" or "destiny," are less damaging to ethics, since whatever our theories, fate and choice seem to be complements in many religious/spiritual sensibilities, rather than alternatives.

Similarly, one might take Rebecca Dresser's characterization of dignity as another example of contextual determination. On her accounting, a strictly non-contextual and normative conception of dignity seems to disappear, subsumed into other moral norms in bioethics such as decisional autonomy, informed consent, and the injunction to do no harm. However, the local discursive context in which a claim to dignity is asserted—for example, a physician-patient relationship—provides dignity with significantly more meaning. In this context, dignity names what scholars in communication studies might recognize as an embodied performative mode: actions and discourses that signal, for example, the need to take seriously the intrinsic value of a person in ways not reducible to adhering to regulations involving informed consent and patient autonomy.

Finally, Celeste Condit emphasizes that the rhetorical functionality of a bioethics argument is as much determined by non-rationally mediated emotions as by the rational force of ideas. When one pays attention to the metaphors that characterize biomedical discourses, there is a world of emotive difference, for example, between saying that a gene is a "trigger" for a disease process and saying that it produces a predisposition for a disease. Though these two descriptions address the same biological function, the feeling produced by use of the term trigger evokes particular negative associations. Differences in the emotional effects of different metaphors depend on the context within which each metaphor is employed, and on the range of associations that each has for a specific deliberative public.

CONCLUSION

We are arguing for a sharper conception of publics as specific communities of attention which by their very nature are constituted by both commonality and diversity of discursive practices and capacities. By extension, a responsible framework for public moral argument requires a commitment to modes of arguing that respect the uniqueness of each deliberative community and tend toward deep consensus by fostering responsiveness to deliberative exchange, revisability, and openness to one's interlocutors.

This argument directs us toward a communicatively constituted conception of social responsibility that provides a framework for public moral argument in bioethics and the conditions of possibility for productive ethical judgments. Attention to context is itself an ethical obligation entailing responsibility for paying attention to the diversity of actually existing publics and the unique modes of meaning-making that inform them. Contextually attentive argument makes possible the process and product of deep consensus on issues of bioethical concern.

On one account, individuals have a responsibility to engage other citizens in the process of decision-making because for each of us to be truly free it is necessary that each individual act responsibly toward others. This vision of responsibility affirms that a vision of what it means to be a human both obligates each of us to act in specific ways, and simultaneously limits the scope of ethical goods that are both desirable and achievable. If it is the case that the biotechnological revolution is going to change what it means to be a human being, a notion of responsibility too firmly rooted in a conception of human nature risks becoming outdated almost as soon as we declare it. Biotechnology harbors possibilities for both expanding the ambit of our biological capacities and, perhaps more significantly, of fundamentally changing our relationship toward human finitude, even including suspending the prospect of death. We might heed the call to an expanded conception of responsibility not rooted in a normative obligation centered on an understanding of human nature, but rather in a communicatively centered conception of "response-ability."

There is an alternative way of locating responsibility that is dependent on neither a static conception of the human subject nor on the promulgation of rules for ideal discourse. This understanding holds that one exercises a form of response-ability in engaging another in speech. It is impossible to deny that to be a subject who speaks is already to be produced by and constituted in relation to another. Thus response-ability moves the seat of responsibility from the individual subject to the exchange of meanings that constitutes subjects. Derrida argues in *The Gift of Death* that responsibility entails a demand that we attend hospitably to the other (Derrida 1996). On this account, to speak or act is already to be positioned by and constituted by the call of the other. Exposure and therefore indebtedness to the other is the condition of possibility for being heard, a sentiment elegantly captured in Michael Hyde's theorization of rhetoric as the "life giving gift of acknowledgement" and as a common "dwelling place" (Hyde 2006). This communicative conception of responsibility suggests that communication studies and bioethics have a good deal to offer one another, if only they are mutually response-able.

NOTES

1. This essay is dedicated to the memory of Ross Smith, who passed away while it was in nascent form. I have tried my best to integrate points that I could decipher from Ross's notes and from our conversations in the pre-drafting stage.
2. Perhaps more importantly, new modes of mediation demand common attention by bioethics and communication scholars, because they themselves imply a bioethical problem, albeit one not directly related to modifications in human biology. McLuhan argues in a chapter in *Understanding Media* entitled "Television, the timid giant" (1994) that the most important effect of new media technologies is that they remake what it means to be human by fundamentally altering the human sensorium. Technology in media studies, as in bioethics implies more than a simple addition to the human, rather media technologies fundamentally change our capacities for sensation, interaction, and action, thus implying a range of questions that have strong resonance with bioethical inquiry.

REFERENCES

Churchill, L. (2011). In the stars or in our genes: The languages of fate and moral responsibility. In this volume: 74–85.

Condit, C. (2011). How we feel with metaphors for genes: Implications for understanding humans and forming genetic policies. In this volume: 123–140.

Dresser, R. (2011). Dignity can be a useful concept in bioethics. In this volume: 45–54.

Derrida, J. (1996) *The gift of death.* Chicago: University of Chicago Press.

Dewey, J. (1927). *The public and its problems*. New York: Henry Holt & Co.

Elliot, C. (2011). An investigative bioethics manifesto. In this volume: 141–153.

Fraser, N. (1990). Rethinking the public sphere: A contribution to the critique of actually existing democracy. *Social Text* 25/26: 56–80.

Hyde, M. (2006). *The life giving gift of acknowledgement: A philosophical and rhetorical inquiry*. West Lafayette, IN: Purdue University Press.

Kalimtzis, K. (2000). *Aristotle on political enmity and disease: An inquiry into stasis*. Albany, NY: State University of New York Press.

Lippmann, W. (2005). *The phantom public*. New York: Transaction Publishers.

McLuhan, M. (1994). *Understanding media: The extensions of man*. Boston: MIT Press.

Moreno, J. (2011). Bioethical deliberation in a democracy. In this volume: 14–24.

Parrot, R. (2011). Responsibility versus "blame" in health communication: Where to draw the lines in romancing the gene. In this volume: 86–122.

Strum, A. (1994). Bibliography of the concept of offentlichkeit. *New German Critique* 61: 161–202.

Warner, M. (2002). *Publics and counterpublics*. New York: Zone Books.

Zarefsky, D. (2011). Arguing about values: The problem of public moral argument. In this volume: 3–13.

Afterword

Nancy M. P. King and Michael J. Hyde

In our introduction we talked about largely instrumental uses of language as a biotechnology. We hope that perusing this collection has suggested some additional, complementary ways of understanding the role of discussion about biomedical technologies. There are several elephants in the room—the moral space—that we are attempting to create for public moral argument, which themselves merit discussion.

The first is the classic Indian elephant, examined by wise men in a dark room (Saxe 1873):

> And so these men of Indostan
> Disputed loud and long,
> Each in his own opinion
> Exceeding stiff and strong,
> Though each was partly in the right,
> And all were in the wrong!

The blind wise men in the famous folktale all had their hands on the elephant. If they had joined their hands together, surely they would have recognized that each was partly in the right. Would they, though, have been any better able to know and describe the true nature of the elephant? That is one of the elephantine challenges posed by a volume such as this: can discourse from two complementary but disparate fields provide a picture of important public issues that is both coherent and meaningful? And can the multiple contemporary sources of public moral argument, each with its disparate and partial viewpoint, come together—with or without the help of scholars in communication and bioethics—to shape a singular portrait of any given bioethics issue, one that scientists, scholars, and publics can in fact talk about together?

The answers to these questions depends on what we think is achieved by talking together (Dink 1999). If we are trying to identify and come to agreement about a common object—that is, if we are trying to describe an elephant—then we might understand conversation about the elephant as perfectly instrumental. But we have already acknowledged that purely instrumental uses of any biotechnology may be incomplete, if the importance of the goal supersedes the value of the process or the relationship that is complementary to it.

Conversation might alternatively be focused not on identifying an object but on reaching an objective, such as formulating public policy. This model might be described as Darwinian: the "best" discourse "wins," as in formal debate or legal argumentation. And a third model might present discourse as negotiation: those engaged in discussion must reach consensus in order to take some action. Public policy-making might be viewed in this way as well. These two models of discourse might thus seem quite relevant to public moral argument about biomedical technology and its place in modern society.

But perhaps there is something else at work in public discussion—something that expands our view of discourse further, by emphasizing its intrinsic value more than its instrumental use. Does talking together have inherent worth apart from the conversation's goals? "Openness to what others have to say, respect for them as sources of intelligibility, a willingness to share our thinking and to risk having it criticized and rejected, a readiness to be moved and changed in response to what others say—perhaps we can recognize these as the virtues which make genuine conversation possible without reducing them to the means to achieve some goal beyond the conversation. Perhaps one could defend this supposition if one could show that conversation with others is an essential condition for having a common and intelligible world about which we can seek to achieve clarity or in which we can pursue our ends" (Dink 1999). This way of thinking about conversation reflects a perspective on autonomy—that key principle in biomedical ethics—that connects it intimately with human moral agency, acknowledging a depth and reflexivity that is perhaps too often flattened and scrubbed pale in bioethics practice. There may indeed be good reason to view public moral argument about biomedical technology as not only having goals like consensus-reaching and policy-making, but also as enacting, supporting, and fostering the nature we share as social human beings. Moreover, the practice of public moral argument surely also both illustrates and teaches responsible decision-making under increasing uncertainty and the fast-changing complexities of biomedical technology.

Recently an important story about genetic research, commerce, and the courts made news all over the country. As a geneticist colleague reported: "Did I tell you I almost was on the network evening news about all of this? I talked with the producer for about an hour and they were actually going to have me on and talk with a reporter. In the end they decided it was just too complex a subject to cover in the 90 seconds they had" (Evans 2010).

If 90 seconds on the national news is the reality of our time, how can bioethics and communication scholars help to foster public moral argument and social responsibility regarding science and health care? The scholars writing herein suggest the need to find, multiply, and reconfigure those 90 seconds at many levels, in multiple relationships, and in many venues, personal, academic, and popular—and to recognize, acknowledge, and support the efforts of others to do the same, wherever they are found. Years ago, the anthropologist-turned-physician Mel Konner, writing about his medical education, reported this advice from a mentor: "Light your corner." (Konner 1987) The essays in this volume suggest to us that we each must take responsibility for lighting our corners—and that, taken together, we can produce enough tiny lights to illuminate a very large moral space indeed: one that is large enough for a great group of wise women and men to see and embrace the whole elephant.

Yet there is a second challenge posed by the elephant in this volume. It comes from the other well-known elephant aphorism: the recognition that some issues for public discussion are extremely difficult or risky to discuss and therefore are often avoided. The subject matter of bioethics falls readily into this category. Imagine the wise men doing their best, not to mischaracterize the elephant, but to miss it entirely. For scholars in bioethics and communication to engage in meaningful conversation with a willing and thoughtful public requires both clarity and moral courage from all participants. Acknowledging the elephant can be arduous, and as risky as it is rewarding, but we have no doubt that it is necessary. We therefore close with the hope that this collection of essays at least begins to reach in these essential directions.

REFERENCES

Dink, M. (1999) What do we hope to achieve in discussion. Convocation address. Annapolis, MD: St. John's College.

Evans, J. P. (2010). Personal communication. Chapel Hill, NC. 2 April 2010.

Konner, M. (1987). *Becoming a doctor: A journey of initiation in medical school.* New York: Viking Press.

Saxe, J. G. (1873). The blind men and the elephant: A Hindoo fable. In *The Poems of John Godfrey Saxe.* Boston: James R Osgood & Co. 77. Available from <http://wordinfo.info/unit/1?letter=b&spage=3>. Accessed 7 June 2011.

Editors

Nancy M. P. King is Professor of Social Sciences and Health Policy at Wake Forest University School of Medicine and Co-Director of the Center for Bioethics, Health, and Society at Wake Forest University.

Michael J. Hyde is The University Distinguished Professor of Communication Ethics, Department of Communication, and Primary Faculty in the Center for Bioethics, Health, and Society, Wake Forest University.

Contributors

Larry R. Churchill, Ph.D., is Ann Geddes Stahlman Professor of Medical Ethics, Center for Biomedical Ethics and Society, Vanderbilt University Medical Center.

Tracey Banks Coan, J.D., is Associate Professor of Legal Writing, Wake Forest University School of Law.

Celeste Condit, Ph.D., is Professor of Speech Communication, University of Georgia.

Christine Nero Coughlin, J.D., is Professor of Legal Writing and Director, Legal Research and Writing Program, Wake Forest University School of Law, and Primary Faculty in the Center for Bioethics, Health, and Society, Wake Forest University.

Rebecca Dresser, J.D., is Daniel Noyes Kirby Professor of Law and Professor of Ethics in Medicine, Washington University St. Louis.

Carl Elliott, M.D., Ph.D., is Professor in the Center for Bioethics, and Professor of Pediatrics and Philosophy, University of Minnesota.

Steven Giles, Ph.D., is Associate Professor of Communication, Wake Forest University.

Eric T. Juengst, Ph.D., is Professor of Social Medicine, and Director, Center for Bioethics, UNC School of Medicine, UNC-Chapel Hill.

Marina Krcmar, Ph.D., is Associate Professor of Communication, Wake Forest University.

Barbara Lentz, J.D., is Associate Professor of Legal Writing, Wake Forest University School of Law.

Christian O. Lunberg, Ph.D., is Assistant Professor of Communication Studies, University of North Carolina, Chapel Hill.

Jonathan Moreno, Ph.D., is David and Lyn Silfen University Professor and Professor of Medical Ethics, History and Sociology of Science, and Philosophy, University of Pennsylvania, and Senior Fellow at the Center for American Progress.

Roxanne Parrott, Ph.D., is Professor of Communication Arts and Sciences, The Pennsylvania State University.

Ross Smith, M.A., was head of the Debate Program at Wake Forest University.

David Zarefsky, Ph.D., is Owen L. Coon Professor of Communication Studies, Emeritus, Northwestern University.

Index